THE ART OF WARGAMING

THE ART OF
WARGAMING

A Guide for
Professionals and Hobbyists

PETER P. PERLA

Naval Institute Press　　■　　Annapolis, Maryland

Library of Congress Cataloging-in-Publication Data

Perla, Peter P.
  The art of wargaming : a guide for professionals and hobbyists /
Peter P. Perla III.
      p.   cm.
  Includes bibliographical references (p.).
  ISBN 0-87021-050-5
  1. War games.   I. Title.   II. Title: Art of war gaming.
U310.P45   1990
355.4'8—dc20   89-28818

9 8 7

To Steven and Sara—may they never see the game become a reality; and to McCarty Little and Frank McHugh—may their work on the first help keep us from the second.

CONTENTS

LIST OF FIGURES

FOREWORD

Admiral Thomas B. Hayward, USN (Ret.)

The battleship is back! Indeed, the four WWII behemoths, now averaging 45 years since commissioning, have returned to active service for what could be another 30 years as the centerpiece of the Battleship Battle Group, perhaps the ultimate in surface action combatant forces of the twentieth century. As I and the OPNAV staff arrived at the decision to reactivate and modernize these magnificent ships in the late 1970s, a question frequently directed our way by the critics—and they were many—was, to what purpose? To what avail? In what scenario? Against what adversary? Advocates of Dr. Peter Perla's thesis on the utility of wargaming will be dismayed to know that none of these questions were fended off with the aid of gaming analyses. Our professional judgment prevailed—and only history will tell us if that was sufficient. I hasten to acknowledge, nevertheless, that OPS analysis/gaming techniques should have been employed as we grappled with this decision, an acknowledgment I readily make at this juncture, particularly after having the advantage of new wisdom invoked by reading *The Art of Wargaming*.

There is a message in this of some import, I would suggest, which is that the pages that follow need to be digested by "operators" and "tacticians" of the naval profession, not just by inquisitive and intent Ph.D.s and theoreticians. Perhaps that explains with some justification why this operator has been asked to set the stage for the benefits that are to come to every reader of this important work. Operators are known for their skepticism of the usefulness of wargames in their insatiable quest for new tactical and strategic options. This one is no exception. The history of combat in the twentieth century provides a plethora of support for such skepticism. Artfully countering this viewpoint, however, *The Art of Wargaming* explores extensively and thoroughly the history, evolution, employment as well as uncertainty of wargaming so as to make this volume one of the most thorough compendiums of wargaming thought available on this controversial subject and a "must read" for the professional.

Wargaming likewise makes the reader acutely aware of the fact that as with most every dimension of warfare, the pace of change in wargaming techniques in the last several decades has been nothing less than awesome. To naval officers of my vintage, who were trained on the checkerboard floor of the Naval War College and weened on the now archaic NEWS, Admiral Nimitz's observations regarding the value and utility of wargaming as it applied to the ultimate execution of the Pacific War constitutes the baseline from which we are prone to judge the marvels and potential of today's wargaming techniques.

In fact, the encroachment of the computer into the world of wargaming has been so stunning and dramatic as to prostitute the term "wargaming" into representing many things to many people, whether it be the hobby game, pol-mil, or the full-spectrum global wargame. Biases become inevitable. Mine reside significantly on the side of focused use of simulation (gaming) as a tactical training tool born of the expediency of real-world exigencies. The speed of encounter confronting today's tactician, brought about by the explosive rate of change in weapons technology has placed the individual combatant in an almost untenable position. Not only has the factor of time been compressed by orders of magnitude in the realm of decision making, but the devastating destructiveness and precision of modern weapons impute a burden for wisdom that staggers the imagination. Yet, wise and accurate the commander must be. And, not just the commander. The burden flows down to the lowest tactical element. The Persian Gulf incident of 1988 in which the USS *Vincennes* launched surface-to-air missiles at a supposedly hostile target only to blow a commercial airliner carrying 290 souls out of the sky is a vivid case in point.

A lesson seems clear. Wargaming, in terms of extensive employment of tactical simulators, is no longer a luxury. It is an essential element of combat team training. It is my conviction that wargaming as a training tool constitutes the overriding value of the wargaming technique, whether tactical or strategic. I would almost go so far as to say the only value. For I contend that using the wargame beyond the training dimension is fraught with flotsam that endangers the utility of the outcome unless managed with great care by the experts. Numerous examples are brought forth by Peter Perla in the pages that follow to support this contention. The number of instances in which "one-game experts" expound mightily of lessons learned that are well off the mark is disquietingly large. Yet, it would be foolhardy and disingenuous to suggest that strategic and pol-mil wargaming are not without value. They are, and must continue—but with due caution as to how the end product is employed.

It is also important to bear in mind that the vast majority of major decision makers in times of crisis and conflict will be those who have seldom if ever participated in a wargame of any magnitude. It is not unimportant to note that a major reason for Admiral Nimitz's admiration for the value of the strategic wargame played during the interwar years was the fact that it had been repeated *more than 300 times* before being put to the test. Therein explains the potential for great advantage today in the employment of the computer-aided wargame, especially as a research tool—with its facile ability to replicate and vary key inputs. Valid lessons are indeed possible from wargames *if* one is fortunate enough to have the time and assets for repetitive play in a variety of scenarios looking for trends and criteria that are consistent and immutable. Even then, the assumptions themselves can be unwittingly flawed, thereby trapping the wargamer in his or her own deception. Happily, even assumptions are now amenable to change and replay with some degree of ease.

The use of simulators as a gaming technique to present indi-

xiv viduals, units, and even battlegroups with a wide variety of tactical scenarios can provide a volume of encounters of sufficient magnitude as to develop, in fact, a highly competent combat team. Procedures, doctrine, and teamwork can be honed to a relatively sharp edge. Repetitive training proves itself important. A degree of experience can be achieved that is not otherwise available from another source. But combat, fear, danger, losses, the unexpected, and fatigue will still be missing—omissions that must never be overlooked.

I recall well when one of the better uses of the wargame as a tactical decision-making training device was instituted by Admiral Ike Kidd in the mid-1970s when he served as Commander in Chief U.S. Atlantic Fleet. Under his astute direction, the Tactical Command Readiness Program was implemented to expose his principal operational flag officers and their staffs to a variety of complex tactical situations, culminating in at least one war game at the Naval War College. The invention of this training program was highly praised. The utility to the commanders involved is unquestioned. Even then, however, the interpretation of its importance as a strategic planning device was significantly overextended by the Washington bureaucracy into the arena of budgetary infighting. Its ultimate absorption into the heated debates over the "Maritime Strategy" of the 1980s was not only unfortunate and highly disruptive to a clear appreciation of the essence of that strategy but was often exploited by those critics of the U.S. Navy's build-up in terms that were not endearing.

The higher the aggregation of the gaming scenario, the more vulnerable the outcome is to misinterpretation and abuse. The Global Wargame at the Naval War College is the most obvious case that comes to mind. Initiated in 1981 on my watch as CNO, the Global Wargame has been repeated annually with increasingly enthusiastic participation and estimates of its value. No doubt these evaluations are justified. But I would suggest that perhaps after it too has been run 300 times or more, one might then be permitted to draw some conclusions with a reasonable degree of certainty of their validity. Which is not to suggest that global wargames should not be continued. They should for the educational value they serve and the opportunity afforded wargame researchers to strive to improve the process.

Thus, the art of wargaming is indeed just that—an art. Dr.

Perla provides the student of wargaming an exceptional book that extends across the full history of wargaming in its many forms and will no doubt provide the reader an understanding of this artful tool significantly beyond the grasp most will have had before turning over its first page. Such has certainly been my experience.

Lastly, Dr. Perla's concern that the cyclical interest in wargaming shows evidence of being on the wane at present deserves our attention and prompt rectification, if so. In some ways, it could be suggested that if the use of tactical simulators is an appropriate classification of a wargaming tool, then it is not likely that wargaming will ever again be on the demise. For simulators are here to stay and will only become more effective, as well as more complex. With the introduction of artificial intelligence, knowledge-based system techniques will be but one example of new dimensions yet to be exploited. On the other hand, if the concern is over the declining interest in wargaming for pol-mil and strategic scenarios, fluctuations in their attractiveness might well be beneficial. In any event, let it be suggested that as long as there are men and women of intellect, inquisitiveness, imagination, and determination—which is virtually assured—the future of wargaming as an essential art to be mastered by some and exposed to many can be reasonably assumed. *The Art of Wargaming* will go a long way toward making it so.

FOREWORD

James F. Dunnigan

Bear with me for a few minutes and I'll make reading this worthwhile. This is being written from the perspective of someone who has designed over a hundred published wargames, plus numerous DoD-related projects and books. I first discovered commercial (Avalon Hill) wargames during the early 1960s while doing my military service. Being nineteen years old and in an artillery unit on its way to Korea, I found the games a compelling means of discovering more about what I was getting into. It was fifteen years later, in 1977, that I discovered what the DoD gaming professionals were up to. My fifteen years of isolation from DoD gaming were devoted to creating commercial games that informed and entertained, in that order. I was asked, at the memorable 1977 Leesburg DoD game-developers conference, what type of game I would produce for DoD game users. My answer was simple. It had to be a game the decision makers could easily use by themselves, with really minimal training, and obtain believable results immediately. Or, as I put it then, "a game the user could play on a computer terminal at home." I haven't changed my attitude, and such a concept has been technically feasible for over ten years. This approach simultaneously attacks several problems most DoD games suffer from: believability, usability, immediacy, and impact.

Let's face it, most DoD gaming results are viewed by many with more than a little disbelief. Part of this disdain can be attributed to the atmosphere in which DoD gaming operates. Much of the decision making that gaming supports is political in more ways than one. The most pervasive "dirty little secret" of DoD gaming is that the games often are there to support decisions and conclusions already made. The tail wags the dog. Fortunately, DoD game developers are largely an intelligent, adaptive, and persistent lot. And the climate is changing. My work had some impact on the changing atmosphere. The three hundred games I published while running SPI reached a large number of military people and many DoD game developers. This had an impact. As one of the few non-DoD game developers invited to speak to DoD audiences, I constantly remind everyone what can be done and describe several ways to do it. Basically, you obtain good games by paying attention to past experience (history) and letting the chips fall where they may. Combat is a dispassionate arbiter of what works and what doesn't. If your games reflect political rather than combat reality, you're likely to find yourself fatally ill-prepared on the battlefield. More users and developers of DoD games are coming to appreciate this. However, current peacetime illusions will always carry more weight than future wartime reality. Unless someone is shooting at you, immediate political demands take precedence over potential military ones. This can change if you actually develop realistic wargames, use them diligently, and widely distribute the results. This was the advantage of commercial wargames, especially the ones that covered current and future wars.

I was shocked, but not particularly surprised, when the Arab-Israeli Wars game we developed during the summer of 1973 proved to be a very prescient predictor of what would come to pass during the actual October War. There were many other incidents like this, particularly with regard to weapons and troop performance. The DoD games are beginning to move into the realm of the real. For example, the Pk (probability of kill) values derived during realistic exercises at the army's National Training Center were found to be only half those used in one of the army's more widely used simulations. The variances are usually larger but are beginning to converge with reality. The historical record is now more frequently consulted for guidance on what should be going on in DoD wargames.

It's fitting that this book has heavier emphasis on naval wargaming. For a number of reasons, naval wargaming has achieved a qualitative and quantitative edge over land and air gaming in this century. This is particularly true with the U.S. Navy, which established a strong gaming tradition between the world wars and carried these traditions forward into the age of high-tech gaming after World War II. Never underestimate tradition and momentum. The U.S. Army has never gotten a full head of steam in the gaming department and the U.S. Air Force is not much better off. There are other reasons why gaming is more likely to thrive at sea. For one thing, naval battlefields are inherently neater than army and air force equivalents. It is also easier for fleets to exercise together and under combat-like conditions. Armies are ponderous and expensive beasts to exercise, and the air force requires realistic ground-force participation in order to run its own large-scale exercises under something approaching wartime conditions. Remember that a U.S. SSN playing tag with another boat is only the push of a button away from combat. In all other particulars, an exercise like that is nearly identical to the wartime article. This makes it easier to keep the wargames honest.

There is a persistent problem with military games and modeling that can be summed up as "garbage in, garbage out." Nothing new here except that historically there has always been a steady drift from reality in the peacetime military. Warfare is a complex process that cannot be easily understood when you can't actually do it. So it's understandable that peacetime preparations, including gaming, will "drift" away from the unknown wartime reality. Contributing to this drift are new weapons and equipment, new tactics and doctrine, and new political situations. Social changes also have an impact: things like economic growth or decline, different partisan political differences, and the replacement of conscription with volunteers. These changes are complicated by changes within potential enemy nations. All of this is further clouded by secrecy. Even if one or more nations were to successfully game out their own and an opponent's wartime prospects, their opponents would be generally ignorant of these game results and their contribution to changes in military policy.

All of this drift from reality creates an atmosphere of uncertainty and surprise when war breaks out. This chaos is an historical constant. Even the U.S. Navy, which did some superb gaming during the 1930s, was often caught short during the first year of World War II. Their list of cataclysmic defeats attests to this. Pearl Harbor and Savo Island are two of the more noticeable examples. While command lapses were partially responsible for these defeats, there were also serious shortcomings in perception that had not been eliminated by the 1930's gaming.

Comparisons between the pre–World War II gaming and what goes on today are an interesting exercise. World War I was a traumatic experience for most participants. Wargaming was then already an established and respected institution, having been steadily developed for over a century in its modern form. World War I had spawned numerous technical and doctrinal changes. The innovations that just got started in 1918 were carried forward in the 1920s and 30s. Twenty years of peace saw the development and deployment of radically new items like mechanized divisions, long-range bombers, and aircraft carriers. More efficient submarines and weapons of all types were introduced. Doctrine and tactics could hardly keep up. With such chaos to work with, it's a wonder the wargamers were able to do anything at all. That wargamers were able to accurately model future combat with useful reliability shows that all the uncertainties mentioned above do not eliminate the usefulness of wargames.

The changes since World War II include completely mechanized armies, ICBMs and nuclear submarines. "Smart" weapons are replacing troops more and more. Electronic devices are everywhere. Wargames will still work amidst all these untried-in-combat innovations only if the gamers play devil's advocate diligently. This means thorough field testing to create realistic wargame parameters. This is where peacetime wargamers traditionally come to grief. There is always pressure, political and otherwise, to cook the data base. Before the Japanese defeat off Midway Island in 1942, Japanese admirals dismissed wargames that showed they could lose their carriers using their current plans. The Japanese admirals went ahead, and lost four carriers and naval superiority in the Pacific.

You trifle with wargame validity at your own risk.

ACKNOWLEDGMENTS

When I started this book, I never expected that writing the acknowledgments would be one of the most difficult tasks. Where to start? Where to stop?

First of all, I want to thank my wife, Jo Ann. Jo Ann not only put up with many evenings and Saturdays without a husband, but she also read the early drafts and made many suggestions for improving the manuscript. Her help went far beyond the call of marital duty, and it and she are deeply appreciated.

I would also like to thank Dot Yufer, who saw fit to assign me to a navy wargaming project as part of my real job, and Phil De Poy, who supported my continued research. That assignment led by circuitous routes to Thomas B. Allen, author of *War Games*, who introduced me to the Naval Institute Press.

During my research and writing I spoke and corresponded with many people who willingly gave up their time to help me find information or avoid saying too many stupid things. I am especially indebted to Mr. John Johnston, who unselfishly shared some of his own research and insight into the history of wargaming. He, too, is writing a book on the subject, and I wish him much success.

Mark Herman, Bill Nichols, and Roger Nord helped im-

xxii   measurably by discussing their own perspectives on gaming and the ideas underlying their work. I am also very grateful to Jim Dunnigan, not only for his foreword, but also for his sage advice to a fledgling author. Mr. Thomas Shaw and The Avalon Hill Game Company were kind enough to help with my research and allow me to reproduce part of their *Jutland* game in the book. Admiral Thomas B. Hayward provided far more than a foreword for the book. His reference to one of my *Naval War College Review* articles in a book review he wrote for *Proceedings* helped convince me that maybe I did have something worth saying.

To help me say it correctly, the staff of the War Gaming Department at the Naval War College provided many helpful insights and comments on the early drafts of the manuscript. Special thanks go to Captain John Heidt and Captain Jerry Gordon, USN, for all their assistance. Anthony Nicolosi and the Naval War College Museum provided several of the photographs. Captain Jay Hurlburt, USN (Ret.), former head of the War Gaming Department, contributed many useful suggestions and a wealth of personal insights.

I must also thank Mr. William Leeson, who provided the translations of several of the critical articles dealing with the Prussian *Kriegsspiel*. Thanks also to the publishers of *National Defense* for their permission to reproduce in chapter 10 some of the material from the February 1987 issue.

Much of Part II is based on work originally done for the U.S. Navy and Rear Admiral "Skip" Armstrong, USN (Ret.). (All of the work has been cleared for public release.) Some of the many naval officers who contributed their thoughts and experiences to the ideas in those chapters include Captains Monroe "Hawk" Smith, Ted Hill, Pat Brisbois, Don Estes, and Bob Gamba; Commanders Dallas Bethea, Larry Stratton, and Jamie Gardner; and Lieutenant Commanders Pat Cassidy, Ray Barrett, and Mike Willmore.

My civilian colleagues at the Center for Naval Analyses included Darryl Branting, Jack Hall, Al Hepp, Ralph Passarelli, and Richard Simon. Dave Dittmer provided valuable comments on the text from his perspective as a naval operations analyst in Washington, out in the field, and at the Naval War College. Thanks also go to Pamela Charles and Ann Fitzgerald, who helped me track down many old books and articles, to Roger Rudy for his help with some of the figures, and to Pat Thorne for her patience **xxiii** with my copying needs.

Finally, I want to thank, or at least try to thank, two very special friends. Michael French is a hobby gamer who has on more than one occasion administered a beating to my ego. He also waded through the first draft of the book and helped smooth out several rough spots. His most significant contribution, however, was the fact that he refrained from killing me for taking up so much of his wife's time.

In many ways Beth French is ultimately responsible for this book's being written in the first place. It was Beth who told Dot Yufer about my background in wargaming, and so helped launch the assignment that, in a sense, is now finally over. Unlike a chemical catalyst, however, Beth did not simply start the ball rolling and then sit back and watch. Without her help with the typing, initial editing, and proofreading of the manuscript, the book would probably still be scattered all over my office. She saved me from my tendency to get carried away with fancy words and convoluted phrases, encouraged me when I thought I would never finish, and kicked me in the tail when it seemed as though I was trying my best to make sure that I didn't finish. I owe more to her than I could ever hope to repay.

THE ART OF WARGAMING

INTRODUCTION

WHY PEOPLE PLAY WARGAMES   1

Games about warfare have probably existed nearly as long as war itself. I suspect that almost everyone who reads this book has heard the story of how modern chess evolved from an ancient Indian game of war in which the pieces represented the various components of the armies of the time: cavalry, elephants, foot soldiers, and so on. Somehow, though, the idea of making a game out of violence, destruction, and bloodshed seems so self-contradictory as to be almost absurd. Yet, despite this fundamental contradiction, or perhaps even because of it, games about war continue to exist and grow in popularity, among both gaming hobbyists and defense professionals.

Indeed, 1987 was in many ways a banner year for those who devise and use wargames. Early in the year, professional gamers from all four military services testified before the Senate Armed Services Committee about the techniques of wargaming and its influence on the development of U.S. military strategy and policy. Over the Fourth of July weekend, the *Origins* convention drew gamers from around the world to the Baltimore Convention Center. And during that same year, two books about wargaming

2   were published. One, a major work called *War Games* by investigative journalist Thomas B. Allen, dealt with many of the same questions asked by the Senate Armed Services Committee. The other, a less-ambitious book called *Pentagon Games* by long-time hobby-game designer and observer of defense affairs John Prados, looked at how military games are designed and used and included three actual games.

In early 1988, the Foreign Service Institute of the Department of State sponsored a two-day conference in Washington about the use of *Serious Games* to answer *Serious Questions*. The conference was attended by more than two hundred military and civilian wargaming professionals, including one of hobby gaming's best-known personalities, Jim Dunnigan, now a consultant to professional wargamers.

With such a long history and high level of current interest, it might be tempting to think that the use of games and gaming in the study of military topics must be a well-developed and well-understood art form, perhaps even approaching the rigor and status of a science. Why, then, another book about wargames and wargaming so soon after its most recent predecessors?

The fact is that wargames and wargaming are consistently misunderstood, denigrated, even denounced, not only by gaming outsiders, but also by gaming proponents and practitioners. This unfortunate situation is a result of a failure to reconcile the fundamental ambiguities of wargaming, to understand the nature of the tool—the game—and of the process of using it—gaming.

Hobby gamers understood more of the nature of the process earlier than their professional counterparts, but because for the most part they were only vaguely familiar with other forms of military analysis and simulation, they could not clearly articulate gaming's value. Professional gamers, those working for the defense establishment, had a much broader perspective, but were apparently too self-conscious of the toy-store image to accept games for what they are. The professional's reluctance to recognize the inherently dual nature of gaming is well reflected by his tendency to insist on using the two-word term *war game* in place of the hobbyist's integrated word *wargame*.

In 1987, a book titled *Team Yankee* gave a fictional account of the experiences of an American army unit in a hypothetical Third World War between the Soviet Union and the United

States. Later that year, Game Designers Workshop published a
hobby wargame of the same name. The introduction to the game
captures many of the reasons why people play wargames.

> The experience of reading Harold Coyle's *Team Yankee* gives a feel
> for the action that modern combat can have. *Team Yankee*, the game,
> takes the concept one step farther and actively involves two players in
> the actions and decisions that Sean Bannon [the book's central charac-
> ter, and commander of combat team Yankee] and others will have to
> make in a succession of modern battles.
>
> Along the way, players will experience first hand the capabilities of
> modern weapons, both U.S. and Soviet, and see their effects and how
> they are used on the modern technological battlefield.
>
> *Team Yankee* is an easy game to learn, and an easy game to play.
> Its lessons, however, are important ones that need to be learned.[1]

These sentiments are well supported by the findings of a survey conducted by the wargaming hobby's premier independent
magazine, *Fire & Movement*. When asked why they played wargames, the most frequent response of survey participants was "to
simulate what they had read."[2] The sentiment is not a new one; it
was eloquently expressed by H. G. Wells in his 1913 book en-
titled *Little Wars*.

> How much better is this amiable miniature than the Real Thing!
> Here is a homeopathic remedy for the imaginative strategist. Here is the
> premeditation, the thrill, the strain of accumulating victory or disas-
> ter—and no smashed or sanguinary bodies, no shattered fine buildings
> nor devastated country sides, no petty cruelties, none of that awful uni-
> versal boredom and embitterment, that tiresome delay or stoppage or
> embarrassment of every gracious, bold, sweet, and charming thing, that
> we who are old enough to remember a real modern war know to be the
> reality of belligerence.[3]

There, in a proverbial nutshell, does Wells capture the essence of why people play wargames.

Ask a hobby gamer why he plays, as *Fire & Movement* did,
and you will get answers like: "I enjoy the intellectual challenge
and competition"; "I am a student of history"; "I like the fellowship and social exchanges that gaming provides." Such statements
are all true, but they are also incomplete. Wells was more honest;
wargamers seek information, understanding, and (be honest
now!) glory.

Jim Dunnigan said that "the most common reason for playing the games is to experience history."[4] He further defined that

4   experience in terms of "being able to massage information in order to see what different shapes the information is capable of taking." Dunnigan, and many of the other leading figures in the present-day hobby world, focuses on the intellectual side of gaming, on the quest for more information and improved understanding. This focus is not surprising when you consider the quantity and depth of the research that underlies most historical board games—the genre most familiar, and owing so much, to Dunnigan and his contemporaries.

Wells, on the other hand, while acknowledging the intellectual aspects of his game, focused on its emotional side. The wargame provides an opportunity for glory without gore and defeat without destruction. By involving the player as an active participant in the events, not merely as a passive observer, wargaming provides a unique learning experience that leads to a deeper and more personal understanding and appreciation of warfare than can be attained by any other method short of actual participation on the field of battle.

Yet, most hobbyists (and nearly all professionals!) are squeamish about the emotional side of gaming. In the first issue of *Moves* magazine, several prominent game designers, critics, and players debated what is known as "The Rommel Syndrome." This syndrome, named after the famous German desert commander and acknowledged master of mobile warfare, was described as follows: "1) A person who plays war games is essentially undergoing all of the rigors of a real commander, less the horror of war, and 2) a person adept at war games will be adept at real war. These two forms are the Rommel Syndrome: the delusion that each wargamer can become his own Rommel and lead his troops to victory."[5] Where those who expressed concern about this "immature" tendency of wargamers to identify too much with their game personas went wrong was in giving gamers too little credit for common sense. Perhaps the most perceptive response to this attitude was that of Redmond Simonsen who argued that trying to force people to be more "mature" in their attitudes about wargaming was an attempt to suppress "one of the prime reasons people participate in hobbies: to have a controlled model (of the uncontrollable real world) in which to act out their fantasies . . . and they do recognize them as fantasies—they just don't want to be constantly reminded of the fact."[6]

Of course, those who play in professional games do so for reasons that are much different from those of their hobby cousins. Although many professional gamers experience the same feelings of intellectual and emotional challenge that the hobbyists do, that is seldom the reason that they are playing. Typically, their motivation is more mundane; playing is part of their job or their education. As such, they often resent the term "playing" and sometimes even the term "game," preferring the euphemisms (and misleading ones at that) of "exercise" and even "interactive simulation."

Hobbyists have suffered from a sense of self-consciousness and defensiveness about their pastime as well. Anyone who has ever been asked the incredulous question "You mean you play games about war?" (often accompanied by a look of borderline revulsion) might be forgiven for a certain reluctance to discuss the hobby with the uninitiated. As a result, wargaming (and the broader category of "adventure gaming") has been a relatively small and closed society about which the general public knows, and cares, very little. In fact, hobby gaming has suffered somewhat from its connection with fantasy games, whose practitioners enjoy an environment rich in swords and sorcery and have sometimes been accused of aberrant behavior.

Fortunately, the last decade has seen some progress in the public's perception of wargaming. In some ways the burgeoning popularity of the microcomputer, and the growing number of wargames available for these machines, may have helped to broaden wargaming's popular base. Why is this fortunate? Well, for hobby gaming the answer is obvious; new players mean new customers and new opponents to play.

It is less obvious that the trend is fortunate for the defense community or even society at large. Yet I believe that this is so. Although war has always been unpleasant, especially for civilians caught in its wake, the dawn of the nuclear age has made the direct effects of war almost incomprehensible. Earlier societies understood the weapons of their time because they lived with them every day. The sword, for example, was functionally little more than a large and elaborate knife, something even the poor and uneducated knew and used. The effects of such weapons, what they could do to human beings, did not need to be studied because they were experienced at close range. The introduction of specialized mechanical devices for killing, especially the ad-

6 vent of firearms, began the process of separating warfare from the day-to-day experience of the average person.

Politicians have said, probably from time immemorial, that "war is too important to be left to the generals." Modern war is too important to be left to the politicians. Even below the nuclear threshold, the destructiveness of modern weapons is difficult for anyone to fully comprehend. But if modern societies are to engage in warfare, no matter how "limited," then the nature of the terrible price to be paid must be understood more deeply and more widely.

If the use of military force is to be avoided when it is inappropriate, and applied judiciously when it is, then everyone, not just generals and politicians, must understand the consequences. No amount of cardboard-and-paper board games or floppy-disk computer games will bring about this popular understanding of war, not alone anyway. Yet, if they are taken seriously enough by their creators and players to be accepted as valuable learning devices, wargames can have an important role in this long and essential educational process.

This is also true for the professional military wargamer. Wargames have traditionally been used to help prepare junior officers to meet the requirements of higher command by giving them a small taste of the problems and opportunities of commanding armies or fleets. New implements of war and concepts for their use have been explored on gaming tables before they ever appeared in the field. This learning process is, if anything, more urgent today than in the past. It is not only the destructiveness of war that has made quantum leaps in the last hundred years; so too have the cost of the tools of war and the speed with which they can work their effects. The choice of the best weapons and the men who will most skillfully employ them is a major concern for the military and the nation. Wargames and wargaming are important tools for helping to sort through such choices.

All that sounds pretty ambitious, and perhaps it is a bit hard to believe, especially if you have never played a serious wargame. That you believe right now is not important; convincing you that it may be true is what this book is all about. Writing this book has been more than just a way for me to fill dark and stormy nights. It is my hope that by explaining the nature of the tool—the wargame—and the process for using it—wargaming—their true value and place will become clear to both the hobbyist and the professional, both practicing and potential. It is also my hope that reading this book will be more than just a way for others to fill their own dark and stormy nights. If this book can convince wargamers—hobbyists and professionals alike—of the need to integrate the lessons learned by all gamers, the future of wargaming can continue to expand its capability to educate its players by involving and entertaining them.

The first step to take, as is so often the case, is into the past. Part I looks at the history of wargaming in both its hobby and professional incarnations. The goal of this foray into the long and fascinating history of wargames is to discover why they were played, what tools the different generations of wargamers devised, what problems they faced, and what progress they made. The answers to these questions will, of course, vary, not only between hobbyist and professional, but also from place to place and time to time. There is a great deal of common ground, and the threads of several important themes wind their way throughout, many of them coming together in the history of wargaming at the U.S. Naval War College.

Based on an understanding of the history of wargaming, Part II isolates the threads and ties them together. It defines some fundamental principles of what wargaming is and what it is not. It describes the subjects that wargaming is useful for exploring, and suggests how wargaming can best be used. This is the tricky part, for too often in the past wargaming's proponents have oversold their wares as a kind of magic answer to all the world's questions. That certainly is not the case nor is it my intent. Nor is wargaming the worthless, misleading dose of snake oil its detractors would have you believe.

Wargaming is part of a larger toolkit of techniques useful for learning about warfare. Part III of the book begins with a discussion of where and how wargaming fits into the larger picture. This final part builds on the historical perspectives and the theoretical principles of the first two parts, both to examine some of the trends in current navy wargaming from this broadened perspective and to project into the future of wargaming. It argues that the war aspects (or realism) stressed by the professionals and the game aspects (or playability) more often stressed by hobbyists need to become better balanced. This cross-fertilization of hobby

8 playability and professional realism can help both types of games improve their ability to educate and enlighten their users through the powerful medium of active and absorbing involvement in the challenge of making "life and death" decisions.

A key element in understanding the past, present, and future history of wargaming and its lessons can be found in the story of its use by the U.S. Navy, and in particular at the U.S. Naval War College. Despite the danger that its author (a naval operations research analyst, after all) may be accused of professional bias, the book contends that the Naval War College, of all U.S. professional gaming organizations, has most consistently and successfully espoused the use of wargaming as both an educational and analytical tool. This has not been so because of any dedication to developing larger and fancier mathematical or mechanical models, though the college has been intimately involved in such developments. It is true because the source of the Naval War College's leadership resides in an underlying, almost subliminal, philosophy of never allowing the tools to dominate the process, of recognizing the ultimately central importance of the human being as both player and warrior. It is a philosophy of balance, perhaps influenced by the arguments of the War College's most famous son, Alfred Thayer Mahan, for a "balanced fleet." Balance is of central importance in dealing with so self-contradictory a subject as wargaming.

WAR GAME OR WARGAME?

If there is a central theme to this book, it is my belief in the overriding necessity of turning a war game into a wargame, of integrating realism and playability in a delicate balancing act designed to achieve a well-understood and well-chosen objective. A wargame must be interesting enough and playable enough to make its players want to suspend their inherent disbelief, and so open their minds to an active learning process. It must also be accurate enough and realistic enough to make sure that the learning that takes place is informative and not misleading.

Wargames revolve around the interplay of human decisions and game events; this active and central involvement of human beings is the characteristic that distinguishes wargames from other types of models and simulations. A wargame's maps, rules, pieces, or computers are only the media through which competing decisions are implemented and judged. Wargames are tools **9** for gaining insights into the dynamics of warfare. They can help players come to a more complete understanding of the sources and motivations underlying the decisions made by historical commanders by placing players in the shoes (or at least the head-quarters) of those commanders and challenging them to do better. Games dealing with current or future military situations can help explore the potential implications of various courses of action, and can raise important questions whose answers can only be found through actual military operations or exercises, or through careful and rigorous mathematical analysis.

Wargames are best used to investigate processes. They can help the hobbyist and the historian better understand the principles and limitations by which military command was or could be exercised under a wide variety of real or hypothetical past, present, or future conditions. They can help the professional commander and staff officer experience decision making under conditions that are difficult or impossible to reproduce in peacetime (such as massed aircraft carrier battle group operations, or full-scale mobilization). Wargames provide an inspiration and incentive for learning the facts of history or current practice, and they can impose a structure and discipline that force game designers, players, and analysts to organize those separate facts into operationally meaningful packages. They can help wargamers explore the feasibility and implications of alternative strategic plans, concepts of operations, or technological innovations. Finally, wargames provide a unique forum for communicating ideas in vivid and memorable ways, and for discussing the validity and applicability of those ideas in a more empirical and less abstract way than the exchange of scholarly papers.

The process of wargaming provides many opportunities for learning. Learning occurs during the research that precedes and accompanies the design of the game, during the play of the game as the participants absorb and act on the information presented to them, and after the game during the inevitable post mortems (or "hot wash-ups"). Whether it involves informal discussions (or arguments) about how one player was successful over his opponent (usually the result of rolling better dice, at least from the loser's perspective) or formal analysis of the structure and play of the game, the hot wash-up is a crucial part of the gaming process.

10   A wargame's structure is built during the game design and development process. The designer must understand the game's objectives and translate those objectives into the infrastructure, information, and mechanics through which those objectives can be met. After this initial design is prepared, the game development process tests and refines it to ensure as far as practicable that the game is complete, valid, and effective at achieving its goals. Although some of the details of the design and development process may differ between hobby and professional games, the fundamental principles remain fairly constant.

The same holds true, if perhaps to a lesser extent, for the play of hobby and professional games. Professional games are played to meet formal educational or training objectives, or to conduct research about specified issues or concepts. Players in professional wargames must understand that their decision-making processes are the key subjects of the games' investigation or instruction. Effective play of professional wargames requires that the participants prepare for their roles by reviewing the issues to be explored during the game and by understanding the real-world systems and concepts they may be called upon to employ. They must play their game roles seriously and diligently. They must also be familiar enough with the mechanics of play to allow the game to flow smoothly and to enable them to interpret the inevitable artificialities that result from the game's compromises with reality. Postgame comments and criticisms by the players are an important element of the process of analyzing the game, learning what it has to teach, and achieving its objectives.

One of professional wargaming's most serious dilemmas is an almost universal tendency to confuse wargaming with systems or operations analysis. Good professional wargame analysis resembles exploratory science or historical research far more closely than it resembles systems or operations analysis, subjects with which many of the people involved in professional gaming may be more familiar. To be of value, wargame analysis must be based not on a complete numerical tabulation of forces and losses during play, but on careful and comprehensive observation of the gaming process. It must include thorough documentation of critical assumptions and decisions, and the rationales for each. Finally, those who use the results of a professional wargame and its analysis need to be aware of the central difference between wargame models and operations research models. Although mathematical models are essential for simulating the occurrence and outcome of game events, their numerical outputs are best regarded as inputs to the gaming process rather than as its results. Wargaming is not, after all, a good tool for producing answers to technical or quantitative questions.

The professional wargamer must remember not only the possible uses of wargaming, but also its potential abuses. In particular, he must remember that, though powerful, wargaming is still only a tool. It is an imperfect mirror of reality, reflecting it best in the decision-making processes of its players. As a result, interpreting the insights derived from wargaming requires special care.

It is also important for the defense community to remember that wargaming is only one of the tools needed to study and learn about defense issues. For the professional, wargaming must be linked with the lessons of exercises, mathematical analyses, history, and current operational experience in a continuous cycle of research that allows each method to contribute what it does best to the ongoing process of understanding reality. Analysis quantifies the physical parameters and processes, wargaming explores human decision making, exercises test human and mechanical abilities to carry out decisions, and history and current operations illustrate what the actual outcomes of military evolutions might look like. Only by integrating the information available from all these techniques can the defense community hope to gain a better, more balanced understanding of the reality of modern warfare.

It is not immediately obvious that the hobby gamer is under this same sort of compulsion to learn from wargaming and to integrate its lessons with those of other tools and processes. Hobby wargamers play wargames because they enjoy the intellectual challenge that games embody, and for myriad other reasons as well. Such reasons may range from the simple joy of competition to the Walter Mitty fantasy of commanding the operations of armies and controlling the destinies of nations.

What many professionals may fail to understand about their hobby counter-parts is that no serious hobby wargamer plays merely to pass the time. Indeed, the amount of time passed in the play of most hobby wargames would imply that gamers either

12 have a pathological need to kill time, or are in pursuit of some higher goal whose attainment is worth the price of the many hours invested. Even the simplest of today's hobby wargames can take four hours to play to a decision. The most complex hobby games can rival not only the weeks involved in the play of the largest professional wargames, but the months spent fighting the actual historical campaigns they represent.

What is it about hobby wargames that inspires such devotion and sacrifice by their players (not to mention their families)? Competition, fellowship, and spending an enjoyable social evening certainly play a part in the attraction, but those can be obtained in sports, or bridge, or any number of other pastimes. What makes wargaming unique is its ability to teach its players something about war and something also about themselves. These are the same characteristics that make professional wargames important research and educational tools. The designers, players, and analysts of hobby wargames have far more in common with their professional counterparts than either group may imagine or care to admit.

The serious designer of hobby wargames will outline his objectives and research his topic as thoroughly and as carefully as his professional counterpart. The serious player will study the history represented by a game and spend hours, days, or even weeks devising a strategy and plan of action for achieving victory. The serious game reviewer (dare I say analyst?), a breed spawned by the wargaming hobby's growing interest in fostering more accurate and more playable games, will dissect a game's research, assumptions, and mechanics and report on his assessment of their validity and their effects on the play of the game. If it can overcome its self-consciousness and prejudice, the professional wargaming community can learn much from hobby wargamers. This book is only the first step in a long and surprisingly important journey to bring those two communities closer together.

PART I

PERSPECTIVES

1

THE BIRTH
OF THE WARGAME

Nobody really knows when or where human beings first used
small objects to represent the maneuvers of warriors on a stylized
piece of terrain. Possibly the first "toy soldiers" were little more
than polished stones, and the first game board no more than a flat
spot in the dirt. Nor do we really know who was responsible for
the first formal rules for moving the objects around the board and
fighting with them. We do not even know why the first wargame
was invented. What we do know is that toys and games based on
warlike subjects existed long before the dawn of written history.
Archaeologists have unearthed sets of miniature soldiers representing ancient Sumerian and Egyptian armies, and games like
chess and Go, primitive wargames at best but still games dealing with military concepts, have been played continuously for
centuries.[1]

Despite our uncertainty about the true origins of wargaming, it seems somehow fitting and at the same time satisfying
to follow the lead of Captain Abe Greenberg of the U.S. Navy
and credit the invention of the first wargame to Sun Tzu, the
Chinese general and military philosopher whose classic *Art of
War* has influenced and enlightened so many readers for so many

16 centuries.[2] (And, incidentally, has also served as the inspiration for the title of this book.)

Greenberg credits Sun Tzu with creating the game known as *Wei Hai* (meaning "encirclement") about five thousand years ago. Little is known about the game or its actual origins, but it appears likely that it was similar to, and probably the original version of, the later Japanese game of *Go*. Like *Go*, *Wei Hai* used a specially designed abstract playing surface upon which each of the contestants maneuvered their armies of colored stones. In keeping with Sun Tzu's philosophy of resorting to the chances of battle only as a last resort, victory went not to the player who could bludgeon his opponent head-on, but to the first player who could outflank his enemy.

In India at about the same time (give or take a thousand years), a four-sided board game known as *Chaturanga* began to become popular among the nobility. Unlike the spare, lean games of the Far East, *Chaturanga* had lots of local color. The playing pieces were not simple smooth stones, but elaborate representations of foot soldiers, chariots, elephants, and cavalry. These pieces maneuvered over a playing board according to a set of fixed rules, but the outcomes of the moves were judged by the roll of dice.[3]

Most writers on the subject seem to agree that modern chess evolved from *Chaturanga*. If recent experience is any indication, the number of players was reduced from four to two because it proved too difficult to find four people willing to invest the time and effort to play the original version. Similarly, the randomness and luck introduced by the dice was probably done away with by a sore loser who blamed his defeats on bad luck rather than inferior skill.

Over the centuries chess grew in popularity until it achieved its current status as one of the world's foremost games. It is the most highly developed and longest-lived game of its type. Those early games were wargames only in a very abstract way. The stylized, almost cartoon-like representation of actual warfare that games like *Chaturanga* and chess employed were almost certainly never intended to be anything more than introductions to the basic principles of military thinking.

The chess-like games require the players to focus on a well-defined objective and to evaluate the abilities of their own and their opponent's forces. They must analyze the strengths and weaknesses of various dispositions, and devise strategies and tactics to overcome the enemy's strength and compensate for their own weakness, and thus achieve the objective. All of these skills are central to military thinking, yet it is a tremendous step from playing chess on a board of sixty-four squares to leading an army on the field of battle, even an ancient Persian army.

The recognition that games like chess were too abstract to be useful as a tool for teaching the finer points of the art of war led in the mid-seventeenth century to variations and complications of the basic chess model to add more and more military detail and flavor. The first game of this new wave to receive enough attention that its name and existence were recorded for history was the "King's Game" or *Koenigspiel,* invented in the German town of Ulm by Christopher Weikhmann in 1664.

Weikhmann's game was based on chess, but employed a larger board and provided each of its players with thirty pieces. Like chess, each of the pieces was named for a character common in the political and military world of its time. There was the inevitable king, a marshal, colonel, captain, and various numbers of lieutenants, chancellors, heralds, knights, couriers, adjutants, bodyguards, halberdiers, and private soldiers. As in chess, each of the pieces had its own peculiar movement capabilities.

In many respects, *Koenigspiel* and other similar games, which came to be known collectively as "military chess" or "war chess," were little more than fancified and overcomplicated versions of their venerable cousin. These games were heavy on what today we might call "chrome" (period color, if you like), but rather light on technical military content. Despite this fact, Weikhmann claimed that his game "was not designed to serve merely as a pastime but that it would furnish anyone who studied it properly a compendium of the most useful military and political principles."[4]

Weikhmann's remarks are perhaps the first recorded instance of a game designer's overselling his wares. Much of the literature dealing with the history of wargaming suggests that games of the "war chess" variety were considered by the players of their time to be valuable "training" devices. It is difficult to accept such evaluations. It seems highly unlikely that any military professional could have believed that such abstract representations of

18   reality could be useful for much more than teaching the basic terminology and principles of warfare.

Although "war chess" and other early games were thus at best rudimentary wargames, the inevitable processes of fiddling with the rules to increase the games' realism slowly led to the introduction of three fundamental concepts central to the future development of wargaming. Although it is not known when each of these ideas first appeared, all three were incorporated into a game invented in 1780 by another German, Dr. C. L. Helwig.[5]

Helwig's game made use of the concept of aggregation, employing a single playing piece to represent a large body of soldiers or organized combat units. It replaced the abstract, bicolored chess board with a multicolored one that represented different types of terrain. Finally, it employed an umpire to supervise the play of the game (and, we might infer, to referee the resulting arguments between players).

Helwig was the Duke of Brunswick's master of pages, and he designed the game to be both entertaining and educational for the group of young noblemen charged to his care. He sought to entice them into thinking about the important military questions of the time and to teach them some of the basic elements of military art and science.

Although it was a major innovation, Helwig's game still retained many of the basic trappings of chess. The playing surface was larger than a chess board, containing 1,666 spaces. But the spaces were still basically the squares of the chess board. Red squares were mountains, and blue squares were lakes or rivers; light green squares represented marshes, and dark green squares symbolized forests; black and white squares were open terrain, and buildings were depicted by half-red squares.

Physically, the playing pieces were essentially chess pawns, but they represented several different types of military units. Each of the two opposing sides deployed about one hundred and twenty pieces: infantry battalions, light and heavy cavalry squadrons, artillery, and even pontoons and pontoniers. Players also could employ about two hundred special pieces to represent fortifications and entrenchments.

A dotted line across the middle of the gameboard separated the opposing sides at start. The combat units had to cross the line

to try to capture the opponent's main fortifications, which were
deployed in opposite corners of the board. The pieces moved in a
manner somewhat similar to chess. Infantry moved in a straight
line, and light cavalry employed a move very like that of a chess
knight, first thrusting forward and then fanning out to a flank.

Farrand Sayre quotes Helwig from a 1781 letter as saying:
"Numbers of military men, profound in the theoretical and practical science of their profession, examined it; . . . they recognized in it a very efficacious means for attracting the attention of
young men destined for military service, creating in them a taste
for the service, and lessening the difficulties of instruction."[6]

Apparently the game enjoyed some success, and its use
spread to France, Austria, and Italy. Over the next half-century it
spawned a number of imitators and many fancified variations. All
of these games, which were put into the class of "military chess"
or "war chess," clearly and unashamedly stressed the game aspect
of wargaming at the expense of the war aspect. They were designed to be a pleasant entertainment for the petty nobility who
made up their principal audience. They were also designed to
provide a little painless and basic education in the terminology
and principles of war as taught at that time. If they were perhaps
overly rigid and formal to the anarchic eyes of today's wargamer,
they were, after all, a product of their times, when war itself
sometimes seemed as rigid and formal as a mathematical proof.
"Military chess" was in many ways an outgrowth of the general
notion prevalent among the military *philosophes* of the time that
much, if not all, of war could be reduced to basic concepts and
formal rules. As McHugh so aptly described it, "War chess re-
sembled rather than simulated warfare. In some ways it might be
considered as having the same relationship to later war games as
the game of *Monopoly* bears to current business games."[7]

THE FIRST NAVAL WARGAME?

About the same time that military chess was the rage on the continent, a Scotsman, who had never been to sea, devised an ingenious method for representing combat actions between sailing
warships. John Clerk used small wooden models to reenact the
great naval battles of history. But mere historical study was not
his goal; he sought no less than a chance to revolutionize the

20   naval tactics of his day. Using his models, he studied the way that ships moved and fought and devised more efficient tactics for their employment.

In the preface to his major work, *An Essay on Naval Tactics, Systematic and Historical,* Clerk wrote the following. "As I never was at sea myself, it has been asked, how I should have been able to acquire any knowledge in naval tactics, *or should have presumed to suggest my opinion and ideas upon that subject.*" [Emphasis added.] His explanation revealed the seriousness and depth of his study: "I had recourse not only to every species of demonstration, by plans and drawings, but also to the use of a number of small models of ships which, when disposed in proper arrangement, gave most correct representations of hostile fleets, extended each in line of battle; and being easily moved and put into any relative position required, and thus permanently seen and well considered, every possible idea of naval system could be discussed without the possibility of any dispute."[8]

Clerk's approach is well described in his own words. "[A]s often as despatches with descriptions of these battles were brought home, it was my practice to make animadversions, and criticize them, by fighting them over and over again, by means of the aforesaid small models of ships, which I constantly carried in my pocket; every table furnishing sea-room sufficient on which to extend and maneuver the opponent fleets at pleasure; and where every naval question, both with respect to situation and movement, even of every individual ship, as well as the fleets themselves, could be animadverted on; . . ."[9] (If, like me, you were unfamiliar with the word animadversion, it is fairly archaic and means a critical and usually censorious remark. Clerk's use of the term is certainly colorful, if perhaps a bit redundant to modern readers.)

By these means Clerk was able to work out the geometry and the mathematics involved in sea battles. He studied the effects of wind and ship maneuvers, the firepower of their weapons and the damage they could inflict. Through his dedicated effort he began to piece together his insights, which he illustrated with detailed drawings. In 1779 he circulated the preliminary results of his work among a few experienced sailors and prominent naval officers. Their generally favorable comments encouraged him to publish a more complete version in 1782. Copies of this version he sent to various influential naval persons, including Admiral Sir George Rodney, who was soon to take a fleet to the West Indies.

Rodney was particularly taken with Clerk's novel approach to breaking an enemy's line of battle, a tactic seldom employed at that time. Later in 1782 Rodney made an opportunity to put the tactic to the test of battle in an engagement against the French fleet of Admiral de Grasse. Rodney credited Clerk's tactics with the British success, which resulted in the annihilation of the French fleet and the capture of de Grasse. The great Lord Nelson himself was to employ variations of Clerk's techniques to even greater effect in 1797 off Cape St. Vincent, and also in his storied and ultimate victory at Trafalgar in 1805.

Clerk had been in the right place with the right tool at the right time. Although it is difficult to call his device a true game (the lack of active opposition makes the term manual simulation seem perhaps more appropriate), it is difficult to believe that many of his experiments did not, in fact, much resemble games in the truest sense. Clerk's technique was similar to the use some of the eighteenth century military experts made of scale models of infantry units to study the basics of battlefield maneuver. Where Clerk had the advantage over his contemporaries, who were concerned with combat on the land, was that the flat tables on which he maneuvered his model fleets could readily represent the surface of the ocean over which the real fleets sailed and fought. The inventors of land wargames had no such luxury; they were stuck with the chess-like board of squares until someone came up with a better way of representing the complications of terrain.

FROM WAR CHESS TO WARGAME

In 1797, a scholar and military author by the name of Georg Venturini designed a new game that was published in the German territory of Schleswig. Venturini's game was based on the style of "war chess," but pushed its development to its most advanced form. Venturini was one of the lesser military *philosophes*, the author of the impressively titled work A *Mathematical System of Applied Tactics and the Science of War Proper*. He published his game under the title *Rules for a New Wargame for the Use of Military Schools*, a book of some sixty pages.[10]

Venturini's gameboard still used the conventional square

22   grid of Helwig's basic system, but expanded its size to about 3,600 such squares, each of which represented one square mile. As with the earlier game, each of the squares was colored to represent the terrain contained within it. Most importantly, the board was no longer a mere abstract piece of imaginary terrain. It represented a portion of one of the most fought-over stretches of real estate on the continent, the border between France and Belgium.

The playing pieces that maneuvered over the simulated ground represented not only infantry and cavalry brigades, the principal combat arms, but also included various supporting arms and equipment. Pieces of different shapes and sizes symbolized bridges, fortifications, supply magazines, artillery, convoys of wagons, and even field bakeries. Venturini's balanced view of representing an army's need for logistic as well as combat elements is one that modern wargame designers and players would do well to emulate.

Although still restricted by the artificialities imposed by the square grid of his map, Venturini attempted to make his forces move more like real men and animals than chess pieces. He even included restrictions on movement during winter months and incorporated the effects of proper support and provisioning of the combat arms.

As its title indicates, Venturini intended his game to be used by military schools, not any budding hobbyists. Its quantum leap in realistic representation of actual military operations required the introduction of complex rules for movement, combat, and logistics. Indeed, the modern hobby gamer who is impressed (or appalled) by the thirty-two- or sixty-four-page rules books of contemporary high-complexity games should ponder the implications of a sixty-page set of eighteenth century game rules. Undoubtedly, the game was plagued by the bane of all detailed manual games, slow play, and Venturini's comment that "one should not call this officer's exercise a *game*" was probably an accurate assessment of its entertainment value.[11]

Venturini's game apparently achieved a certain amount of popularity with military-minded men in Germany, Austria, and Italy, but it was not without its detractors (a common theme among virtually all wargames). General von der Goltz, an eminent Prussian soldier of the latter nineteenth century, commented wryly that it "resembles very closely the game of 'poste et

de voyage' . . . in which, upon making an unlucky throw of the   **23**
dice, one tumbles into a swamp, or breaks an axletree, or experiences some other such mishap. . . . This war game is a bad product of the refined military education of the period, which had
piled up so many difficulties that it was incapable of taking a step
in advance."[12]

But even as Venturini advanced the square-grid-based wargame to its highest pre–twentieth century form, new developments were on the horizon, portending a revolution in wargaming of immense proportions. The key ingredient, and one
we today easily take for granted, was the introduction of accurate maps.

As early as 1730, a Dutch engineer had devised a means of
using contour curves to represent the bottoms of rivers. By the
end of the century, a version of that technique began to be applied to maps of the countryside. These new maps were used by
Napoleon to great advantage in his conquest of much of Europe.
The new maps and Napoleon's revolution in the art of war provided the motive and opportunity for breaking away from the
confines of war chess and its abstractions. As more realistic models
of the ground became available, new ways of employing them
were applied to the wargame. The watershed, perhaps not surprisingly, came in Germany.

VON REISSWITZ: WARGAMES TAKEN SERIOUSLY

As the nineteenth century broke over a war-torn Europe, the
pieces of the wargaming puzzle lay strewn around the continent.
The process of fitting those pieces together was begun by a Baron
von Reisswitz, who was not himself a soldier, but was a civilian
war counselor (Herr Kriegs-und-Domanenrath) to the Prussian
court at Breslau.[13] Von Reisswitz discarded the game board of war
chess in favor of a sand table in which actual terrain could be
modeled in relief. The playing pieces were made out of wood,
cut to scale to represent the frontages taken up by the military
units of the time. Symbols representing the different types of
units were pasted to the blocks.

By a stroke of good luck, von Reisswitz and his game had
come into contact with a Prussian officer by the name of von
Reiche. The latter just happened to be the captain of cadets at the
Berlin garrison, and in 1811 was responsible for instructing the

24 Princes Friedrich and Wilhelm (later to become no less than Kaiser Wilhelm I) in the art of fortification. He happened to mention the game to his young students, who promptly petitioned their governor, Oberst (colonel) von Pirch II, to arrange for a demonstration at the castle in Berlin where the princes lived.

Von Reisswitz, von Reiche, and another young officer of von Reiche's acquaintance (2nd Lieutenant, and later General, von Wussow) were invited to arrange such a demonstration, and the princes enjoyed the game so much that they wanted to play it again. Von Pirch, who had watched the play of the game with some interest, agreed and also allowed them to tell their father, King Friedrich Wilhelm III, about their experience. The king was fascinated by the idea of this new, much more accurate representation of war, and advised von Pirch that he would himself like the opportunity to witness a demonstration of the game.

Imagine for a moment what von Reisswitz must have felt like when he heard this news! But after his first flush of delight, von Reisswitz decided that the sand table on which he currently played the game was not really appropriate to so great an occasion. He thereupon began the time-honored practice of polishing up the physical presentation of the game before attempting to "sell" it to the boss.

He also nearly blew it by taking so long! It was not until the following year that von Reisswitz finished constructing his fancy new portable system. Although the king had almost forgotten about the game by that time, he agreed to a demonstration, and "was not a little astonished to see something in the shape of a massive piece of furniture arrive."[14]

The new apparatus "was in the shape of a large table open at the top for the terrain pieces to fit into. The terrain pieces were three to four inches square, and the overall area was at least six feet square. The small squares could be re-arranged so that a multiplicity of landscape was possible. The terrain was made in plaster and was coloured to show roads, villages, swamps, rivers, etc. In addition there were dividers for measuring distances, rulers, small boxes for placing over areas so that troops who were unobserved might make surprise attacks, and written rules which were at this stage not yet in their fuller form. The pieces to represent the troops were made of porcelain." The play was driven by a "General Idea" or scenario that described the situation in which the players found themselves at the start of play.

At this stage in its development von Reisswitz's game rules dealt only with the movement of the troops, and not with combat. The results of engagements were worked out by the players themselves. By all accounts one of the players, the king himself, was very enthusiastic, keeping his family up well into the night to play.

By the early 1820s, however, the novelty of the game had worn thin, and the relative peace on the continent contributed to a general decline in interest. Undoubtedly, the unhandiness of the physical apparatus also played a part in this flagging of enthusiasm. Just when it seemed that von Reisswitz's game would follow its predecessors into the closet of history, the elder man's son introduced a new and much improved adaptation of his father's ideas, and one destined to direct the course of much of wargaming's subsequent history.

Lieutenant George Heinrich Rudolph Johann von Reisswitz was a first lieutenant of artillery in the Prussian Guard when he developed and introduced his revised version of his father's game in 1824. The most obvious difference between the two games was the scene of action; gone was the old sand table, replaced by detailed topographic maps drawn to the scale of 1:8000 (roughly eight inches to one mile). In addition, with characteristic Prussian thoroughness, the younger von Reisswitz attempted to codify actual military experience and introduced the details of real-life military operations lacking in his father's game. In particular, he quantified the effects of combat so that results of engagements were calculated rather than discussed.

The game was thus implemented with a detailed set of rules covering virtually every contingency of operations of units up to the size of divisions and corps. These rules, developed by von Reisswitz and several of his officer friends, he finally published in 1824 as *Instructions for the Representation of Tactical Maneuvers under the Guise of a Wargame (Anleitung zur Darstelling militarische manover mit dem Apparat des Kriegsspiels)*.

The publication of this truly seminal work in the history of wargaming was once again due in a large measure to the intervention of Prince Wilhelm, who after an evening's play advised von Reisswitz and his friends that he intended to recommend the game to the king, and to the chief of the general staff, von Muffling. Such a warm reception seems to have given von Reisswitz the encouragement he needed to publish the game,

26   much as the earlier attention of the royal family had encouraged his father to jazz up his apparatus.

True to his word, the prince soon arranged for von Reisswitz and company to present their wares to von Muffling. Like many superior officers before and since, the old general was skeptical, if not disdainful, and probably a bit put out by the fact that he had to endure such silliness because of a royal edict by the young and headstrong prince. One of von Reisswitz's companions on that day was a young officer named Dannhauer, who would later attain general's rank. He described the scene as follows.

On our arrival we found the General surrounded by the General Staff officers.

"Gentlemen," the General announced, "Herr von Reisswitz is going to show us something new."

Reisswitz was not abashed by the somewhat lukewarm introduction. He calmly set out his Kriegsspiel map.

With some surprise the General said, "You mean we are to play for an hour on a map! Very well. Show us a division with the troops."

"May I ask Your Excellency," replied Reisswitz, "to provide us with general and special ideas for a manoeuvre, and choose two officers to be the commanders for both sides. Also it is important that we only give each commander in the special idea the information he would have in reality."

The General seemed rather astonished at the whole thing, but began to write out the necessary idea.

We were allocated as troop leaders to both sides, and the game began. One can honestly say that the old gentleman, so cool towards the idea at the beginning, became more and more interested as the game went on, until at the end he exclaimed, "This is not a game! This is training for war! I must recommend it to the whole army."[15]

Unlike many later officers in positions of authority, von Muffling had kept his mind open enough to recognize the potential value in von Reisswitz's fledgling game. The general wasted little time in making good on his pledge to recommend the game to the entire army by publishing the following notice in the *Militar Wochenblatt*, no. 402.

There have already been a number of previous attempts to represent warfare in such a way as to provide both instruction and entertainment. These attempts have been given the name "Kriegsspiel." They have usually presented many kinds of difficulties in the execution, and they have always left a large gap between the serious business of warfare and the more frivolous demands of a game.

It is noticeable that up till now it has only been non-military personnel who have occupied themselves with the wargame invention, and
the resulting incomplete ideas of warfare, and its incomplete imitations
have never seriously been able to claim the attention of trained officers.

At last, after years of trial, insight, and perseverance, an officer has
pursued the topic begun by his father, the Reigerungsrat von Reisswitz,
and has so much extended it that warfare can actually be represented in
a simple and lively way.

Anyone who understands those things which bear on leadership in
battle is able to take part immediately in the game as a commander of a
large or small unit, even if he has had no previous knowledge of the
game or has never seen it before.

The execution of good plans on realistic terrain, and the ability
which the game offers of presenting a multiplicity of situations, makes it
continually instructive.

I will gladly, by all means in my power, assist in seeing the number
of available copies augmented.

If the 1st Lieutenant von Reisswitz has already found reward for
his efforts through the approval of princes of the Royal Household, the
Army War Ministry and high ranking officers who have come to know
of his efforts, the further distribution and knowledge of the game will
earn him the thanks of the whole army.[16]

Pretty impressive. Clearly von Reisswitz had in fact done
something new. His was no longer a game about war or a war
game, but was finally the integrated, balanced Kriegsspiel, capable
of fusing the "serious business of warfare and the more frivolous
demands of a game." Perhaps the best way of describing the
means by which he achieved this long-elusive end is to let him do
so in his own words:

The Kriegsspiel has the aim of representing that moment in warfare when the strategic object can only be realized by an attack.

A general idea of the game is given in the following short
description.

No great preparation is required for the game. Any room can
be used.

Three players are necessary, one of whom will give an idea for the
game and keep a tally of losses. More can take part, and any extra players are allocated equally to both sides.

Ideally four to six players take part. Then one player on each side
becomes the commander-in-chief and the others are allocated command of individual units. One needs to know a few things about the use
of the equipment, but no special practise is needed. Any officer familiar
with the arrangements for the different troop types can take part at once.

A good map in $1:8,000 - 1:12,000$ scale, which gives good terrain
detail can be used.

28   The troops are represented by small metal strips. These are marked up in the same way as they are found on battle maps. They are moved on the map with consideration of time and distance.

Time in the introductory phase and in the course of the battle is divided into two minute intervals. These intervals are called moves. Anything can take place in one move which in reality could happen in two minutes, both as far as movement, fire effect, and hand-to-hand attacks are concerned, and as far as experience and study would indicate.

The troop symbols are so arranged as to give the correct frontage for troops in line and also for troops in march or attack column—the first is given by the length, and the second by the breadth of the blocks.

All movements or positions of the enemy which would remain concealed in reality are similarly undisclosed in the game. The troops in such cases are not put on the map, but the player (umpire) who designed the manoeuvre and who controls the game records their positions. As soon as they reach some point where they could be seen by their opponents they are placed on the map.

To simplify the game, and to affect the players, as in reality, by considerations of good luck or bad luck in the outcome of battles, the results based on experience, for the effect of fire weapons in good and bad circumstances, are stuck on to dice and determine the losses. The attacks with hand-to-hand weapons are similarly noted on the dice so that equal or unequal strengths of forces can be considered. So, for instance, one side might have two or three chances to the enemy's one.

The strength of forces, however, is not only a question of the numbers of troops, and the advantages which might come from favourable terrain, flank attacks, etc., are also taken into consideration.

The object is for the player who has good luck to seize and use it, and for he who has the misfortune to meet with bad luck to take those correct measures which in reality would be required.

The equipment consists of the following:
1. Troops for each side—26 battalions, 40 squadrons, 12 batteries, 1 pontoon train.
2. Rulers and dividers for finding the correct march and firing distances.
3. Dice for deciding fire effect, and the results of hand-to-hand attacks.
4. A small book of six chapters containing the introduction to the use of the equipment and the rules.
5. A map covering 4 square miles (Rheinl.) 1:8,000.

The whole equipment is contained in a mahogany box 10 inches long and 6 inches wide, so taking up little room.[17]

Unfortunately, von Reisswitz could not package the key ingredient of his successful demonstration for von Muffling in the small mahogany box. Almost certainly much of the favorable impression the game made on the old general was due to the ease and smoothness with which von Reisswitz and his cadre of experienced gamers managed to guide the inexperienced staff officers through the play of the game. Although the demonstration convinced von Muffling that anyone with experience in actual military operations could play the game immediately without the need to learn an elaborate system of rules, von Reisswitz's description of the equipment and instructions gives the lie to that assessment. As with any new tool, those unfamiliar with its effective use had to learn its capabilities and limitations, and becoming skilled in its employment undoubtedly required study and practice. These, in turn, depended on devoting a certain amount of time to the game.

A junior officer's time, however, is a commodity of which his seniors often are jealously possessive. Just as von Muffling had at first been cool to von Reisswitz despite (because of?) Prince Wilhelm's recommendation, many senior field officers were hostile to von Reisswitz's game despite or because of von Muffling's endorsement. Many of those officers became converts, just as the old general had; many more were not as open-minded as the distinguished chief of the General Staff and simply refused to accept or even evaluate the game as a useful device. Indeed, the criticism most often levied against Kriegsspiel was that allowing junior officers even a simulated taste of commanding forces beyond the prerogatives of their rank would make them impatient with the day-to-day routine of company-level duty. This, in turn, would cause them to become less conscientious in performing their normal tasks.[18]

Despite arguments, and even evidence, to the contrary, the debates between proponents and detractors of von Reisswitz's Kriegsspiel continued throughout its early years. Von Reisswitz himself soon ran afoul of some of his detractors, who transferred him to Torgau, site of a famous battle of Frederick the Great's, but at that time nothing more than an obscure provincial border fortress far removed from the glamours of royal courts and all-night Kriegsspiel parties.

Von Reisswitz clearly resented his transfer, which he ascribed principally to jealousy, and began to brood about the injustice of it. Cut off from his circle of friends and from any real ability to develop his game further, von Reisswitz soon became

30 despondent. Tragically, in 1827, his meteoric rise to fame and notoriety ended with an equally sudden and dramatic suicide.

THE HEYDAY OF KRIEGSSPIEL

The game von Reisswitz left behind proved more resilient than its inventor. Despite the forces of conservatism and reaction, Kriegsspiel grew in popularity among the future leaders of the late nineteenth-century Prussian war machine. Even the demi-god himself, Helmuth von Moltke, was an avid player of the game as early as 1828.

Through the middle years of the nineteenth century, Kriegs-spiel continued to attract the attention of new adherents in the Prussian military, but spread little outside that country. Inevitably, as more and more officers played the game, their different experiences and perspectives led them to introduce changes in von Reisswitz's original rules. During that period many officers published extensions and modifications of Kriegsspiel's basic ideas, one of the most prominent being the work of von Tschischwitz. Particular dissatisfaction seemed to center around the complex and rigid nature of many of the original rules, and also on the role of the umpire.

Then, as today, a focus of much of the debate and revision was the method used to resolve the outcomes of combat. Throughout the early growth of the game, the usual method of calculating outcomes evolved into a particular pattern, probably given its principal form by the Prussian Captain Naumann, who published his rules in 1877 under the title of *Das Regiments Kriegsspiel*. "Captain Naumann originated the idea of selecting a standard case and deducing from it, by applying a suitable multiplier, the result to be expected in any particular case."[19] In this system, a particular, recurring, basic combat event was defined as the standard. For example, one minute of fire of a body of infantry in a particular formation could be expected to produce a certain amount of casualties on an opposing force. This became the standard result, but was adjusted by a series of multipliers to reflect the particular circumstances of the engagement.

The complications, and computations, were clearly legion in such a system. Over time, this approach spread its tentacles throughout the game, as new standards and new multipliers were added for new weapons and tactics and situations. As a result of this massive compilation of details, the role of the umpire evolved into virtually that of a computer. Tables of logarithms, to speed up the process of multiplication, became nearly as essential to wargaming as maps. The umpire's main contribution, besides his arithmetic skills, became deciding just what set of multipliers to employ in a given set of circumstances.

The system of increasing detail and pseudo-exactitude, whose origins lay in the desire to make the game more realistic, detracted from the very realism it had hoped to improve. Play became tedious and slow, dominated by the calculation of losses, which would often prove to have little real effect on the decisions of the players.

"FREE" VERSUS "RIGID" KRIEGSSPIEL

The stunning Prussian success in the "Six Weeks War" against Austria in 1866 was followed only a few years later by their even more stunning defeat of the French. In the aftermath of the 1870–1871 Franco-Prussian War, European and world military opinion suddenly became enamored of things German, including Kriegsspiel, to the use of which many experts attributed the German victories. At the same time, the tremendous practical experience of battles and war, which had hitherto been somewhat lacking in the practitioners of Kriegsspiel, now made its weight felt. The reaction that had been so long in coming now struck with the force of an exploding shell.

In 1876, Colonel (later General) Jules von Verdy du Vernois opened the ball with a slim volume that did little more than briefly describe a simple, but somewhat revolutionary, idea and give an example of its use. Verdy was one of the foremost military writers of his day, and was especially influential in the realm of troop training. He was dismayed by the fact that wargaming was still having a difficult time establishing itself as an important training and educational device, despite its obvious advantages.

For von Verdy the reason for wargaming's lack of popularity lay "in the numerous difficulties that beginners run against in handling tables, calculating losses, and the like."[20] He argued strongly that "it would add to the usefulness of the game to be rid of these numerous rules and tables."[21]

His proposed solution lay in the technique used by von Moltke during the latter's staff rides.

32   In the excursions of the General Staff, the military results that are sought are the same as in the War Game. . . . In no case do they decide upon the success of an operation by the cast of the die, but each time the umpire decides according to his own views. It is not thought necessary to decide in every case the effects of fire, shock, etc. All that is necessary is to reach the general result, to determine if a body of troops has had great losses, if it has been so badly broken that its power of resistance has been sensibly diminished. Therefore, the means employed on the ground in these staff excursions can be easily applied in the lecture-room with maps. . . .

The mode adopted is to place on a map all the troops as soon as they are in contact. In the staff excursion they require an officer on the ground itself, to answer such a question as: "What would you do if you saw a hostile column coming suddenly from yonder village?" Very good officers, who in real presence of the enemy would see at once the proper thing to do, are often embarrassed and cannot answer. This is because the imagination is not developed to the same extent with everybody. This lack is filled to a great extent by the War Game, and by the placing of the blocks.

The War Game is then used in three different ways, with more or less extension, according to the end in view: 1. The strategic game, on a comparatively small map, showing the larger operations, without tactical details. 2. The game with dice, rules, and computations. 3. The simple method on a large-scale map, without rules, tables of losses, or dice.

Every body of officers should use the system best adapted to its end. Whatever the method, its value as a military study will depend largely upon the degree of ability possessed by the director.[22]

Here, in a nutshell, von Verdy states the case for what was to become known as "free Kriegsspiel," in contrast to the "rigid Kriegsspiel" of von Reisswitz and its descendants. The essence of von Verdy's approach can be described as the transformation of the umpire from computer to "God." But he was not to be a capricious god, but a conscientious one who would explain his actions and assessments after the game.

Although enthusiastically accepted and applied by those who found the rigid games too complex and boring, free Kriegsspiel was obviously not without its own problems and detractors. The contention between the proponents of rigid and free Kriegsspiel brought into sharper focus the inherent tension between realism and playability. More importantly, however, it revealed the fact that the lack of realism could result in a lack of playability, just as a lack of playability could lead to a shortage of realism.

The late nineteenth and early twentieth centuries thus began to see increased efforts to achieve some sort of balance between the false realism of rigid Kriegsspiel and the false playability of free Kriegsspiel. "We find the writers on Free Kriegsspiel occasionally admitting that rules and tables may sometimes be useful; and on the other hand, the advocates of Rigid Kriegsspiel generally begin with the statement that these rules and tables are intended to be merely of an advisory character and that the director should proceed without them whenever his personal knowledge suffices for the occasion. The distinction between the two systems is, indeed, largely one of degree; yet it is sufficiently marked to justify its recognition."[23]

Despite such attempts at balancing rigor and speed of play, turn-of-the-century wargames were at best a mixed blessing. Sayre, whose book *Map Maneuvers and Tactical Rides* gives a good account of the state of the art in the early 1900s, was of the anti-game school, arguing for the use of the term map maneuver, and for downplaying the competitive "game" aspects of the exercise.

Although map maneuvers owe their origin to a game, the game feature is no longer an important element of them. A predominance of the game idea has always been an obstacle to the proper development of these exercises as a means of military instruction and training. Through the influence of the game idea, the exercises have often been carried farther than was necessary or profitable. In order to ascertain which of two parties of players would be the winners, it was necessary to push an engagement through to a decision; and in order that no injustice might be done to either side, computations of losses were made, and accurate records kept, by a definite and uniform system. The dice, tables, and rules served to secure fairness to the players and to clear the director of the suspicion of bias; but when the idea that the map maneuver is a sort of game—in which one merely plays to win—is set aside, these considerations lose their importance.[24]

Despite the solid good sense of much of his work, here Sayre errs. He is right to argue that it is the learning and not the winning that is important when games are used as educational or training aids; he is wrong to assume that such learning does not need to worry about the possible biases of the directors. Biased direction can teach invalid, and potentially deadly, lessons. Solid research and mathematical modeling, embodied imperfectly per-

34 haps in the games of Sayre's time, are important elements of the educational process, not merely devices to help assure competitive fairness.

Yet, it is difficult to dismiss all of what Sayre has to say on this subject, for he clearly perceived the central problem of achieving realism in any wargame that attempts to deal with actual combat. "A few hundred years ago, battles were fought by common consent of both armies, and the art of war was limited to the battlefield. But at the present day the most important movements of troops generally takes place out of sight of the enemy, and these we can feel sure of representing faithfully on a map; but we cannot feel so sure of representing correctly the latter stages of combat, for no one can tell with certainty the manner in which future battles will be fought."[25] It is this very lack of certainty, however, that makes wargaming so important. We may never know the right answers, but gaming can sometimes help us learn to ask the right questions.

Unfortunately for Europe, too few military men and politicians were asking themselves the right questions in the first decade of the twentieth century. As the storm clouds of World War I gathered on the horizon, one of the great writers in English history, and also one of that society's most fervent pacifists, brought a new perspective to the subject of games about war, and in so doing gave the principal impetus to the development of something new, a serious amateur wargaming hobby.

H. G. WELLS AND MODERN HOBBY WARGAMING

Today, Herbert George Wells is probably best remembered as a writer of science fiction; *The Time Machine* and *War of the Worlds* are probably his most well-known works. Yet, Wells was much more. He was an historian, publishing his monumental *Outline of History* in the 1920s. He was a well-known pacifist. He was also interested in games, publishing *Floor Games* in 1912 and, most importantly for our story, *Little Wars* in 1913.

Little Wars described a system for playing battle games using toy soldiers made of lead and a spring-loaded cannon that fired a wooden projectile capable of knocking the men over. The battlefield was constructed of model houses, miniature trees, and various other ingenious devices to give the toy soldiers an appealing and interesting country over which they could engage in mortal **35** combat.

Some form of toy-soldier warfare had probably existed as long as the toys, but Wells's system was the first to receive widespread attention, among other reasons being that while his basic ideas were quite simple and easy to implement, they were also easily extended and adapted to more and more complex situations. Play proceeded in turns during which each of the two opposing sides could maneuver and fight. Each of his three combat arms—infantry, cavalry, and artillery—were imbued with certain unique abilities to move and engage in fire or "shock" combat. Fire, carried out in the basic system only by the guns, simply consisted of firing a set number of shots from each eligible cannon each turn. Any soldiers knocked over by the cannon fire were considered killed and removed from play. When soldiers from one side moved close to soldiers from the other side, a close combat ensued in which a certain number of pieces from each side would be eliminated. Special provisions were made for heavily outnumbered or surrounded forces to surrender rather than fight to the death, and for the escort and release of prisoners.

Although Wells intended his game to be nothing more than an amusement and not something of use to the professional military, at the urging of professional acquaintances he included in an appendix to *Little Wars* his thoughts on how some of his principles might be applied to the professional kriegsspiel of the time. Wells's approach was, in some ways, a return to the representative physical model used by the elder von Reisswitz. But Wells went one step further. He opposed the standard professional practice in which the results of actions were either decided by an umpire based on his own experience and prejudices (the "free kriegsspiel" described earlier) or calculated from extensive charts and tables ("rigid kriegsspiel"). Instead, Wells argued that the results of combat "should be by actual gun- and rifle-fire [using the toy cannon] and not by computation. Things should happen and not be decided."[26] Thus Wells advocated the use of physical representations not only for terrain but for combat as well.

Wells's ideas on replacing calculation and arbitrary judgment with a physical model of combat, intriguing and appealing as they are from a philosophical viewpoint, have never had much

36   of an effect, especially on professional wargames. Some of the resistance to using something like his spring-loaded cannon in a professional game was undoubtedly because of its too obviously toylike (not to mention undignified) nature. But it was also impractical to represent an action much larger than a small skirmish with the equipment and approach taken by Wells. For larger battles there appeared to be no substitute for the map, and no opportunity to represent combat physically.

Despite the fact that one of his key ideas never achieved a great degree of popularity, even among hobbyists, Wells is usually considered the father of modern wargaming with miniatures. Indeed, for many years the gaming hobby annually awarded the "H. G. Wells Awards" to companies and individuals who had made a significant contribution to the miniatures gaming hobby during the preceding year. The publication of *Little Wars*, coupled with the ready availability and affordability of the mass-produced soldiers and cannon needed to play it, finally made wargaming a hobby that anyone, not just the well-to-do, could pursue.

Although the most prominent of the early twentieth century's amateur wargamers, Wells was by no means the only well-known figure to enter the field. In fact, several years before the publication of *Little Wars*, Fred T. Jane, the editor and driving force behind the long-lived annual *Jane's Fighting Ships* (originally *All the World's Fighting Ships*), published a set of rules for playing wargames with cast miniature warships. Interestingly, Jane's game employed the same notion of having things happen rather than just making calculations.

Jane certainly did not view his game as a toy or as part of any "gaming hobby." Instead, he and the many proponents of his game saw it as "a most instructive lesson in the capabilities of different types of ships to withstand or carry on attacks. . . . In short, it puts very fairly before the players the actual problems which would face them were they commanding squadrons in times of war, and if played in seriousness cannot fail to instruct them."[27] Jane himself also undoubtedly saw it as a means to promote his books, which were used as "the textbook of the game," and as a device to allow him to gain access to "naval officers specially selected in all navies" whose opinions served as a source for his method of evaluating the results of combat.[28]

Play of the game used ship models, which were accurately scaled and unmarked. These models had "to be recognized by their opponents just as they would have to be in real war."[29] They were maneuvered on a large playing board marked off in large squares representing 2,000 yards and smaller ones of 100 yards. As each move of the game represented one minute, moving a single small square was equivalent to making a speed of three knots, and so moving five squares was the equivalent of fifteen knots. Diagonal moves required the use of a measuring device, and in especially complex games detailed turning circles were employed.[30]

Ships were maneuvered on the board under the overall direction of the opposing admirals and usually with a single player as captain of each ship. "The admirals, till fire is opened, are allowed to give any directions they please to their captains; after fire is opened they may transmit signals only through the umpires and at the discretion of these, and each captain has to think for himself and carry out his orders as best he can. As the umpires rarely allow anything save the simplest signals to be made in battle, the admiral who lays down his orders clearly beforehand is in much the same position as one who does so in real war."[31]

To evaluate the effects of fire, each ship was "supplied [with] a large number of similar drawings of her on thin card. These drawings, elevation and plan, are divided into sections equivalent to 25 ft. lengths."[32] Each section was rated for armor protection using the scheme defined in All the World's Fighting Ships. The effect of gunfire depended on the caliber and penetration of the firing weapon relative to the protection of the region in which hits occurred.

Jane's method of determining hits was perhaps the most unique element of the game. It used the differently sized drawings of target ships as a function of range or, in an alternative method, it was based "on the combined speeds of the ship firing and the ship fired at, i.e., whether they are relatively stationary or shifting bearings rapidly."[33] Firing itself was "done with a striker, consisting of a strip of thin wood about 15 in. long, with an enlarged end, somewhere near the centre of which a short pin point projects."[34] The selected target view was laid on the table and the firing captain struck at it with his striker, the resulting hole in the target indicating if and where the ship was hit. The number of

38   strikes was determined by the rate of fire of the particular weapon firing.

This approach, so similar in spirit to that of Wells, "seems at first sight crude, and it is only after having used it for some time that its suitability is appreciated. In no two strikers is the position of the pinpoint the same, and the expanded end of the striker is of such size that a large area of the target is covered, so that a sufficiently large margin of chance prevails. It has been found that all the attempts to secure accuracy which different players adopt end, as a rule, in failure, and no matter whether a short strike or a long strike is made, the effect is generally the same. As a rule, moreover, the accuracy of firing gets worse as the excitement increases. It will thus be understood that the strikers do fairly imitate the conditions of actual shooting."[35]

Jane also included rules for the strategic maneuvers of fleets prior to contact, and these rules seem very similar to those employed at the U.S. Naval War College, described in the next chapter. Although Jane managed to raise the ire of many in the U.S. Navy by describing a hypothetical naval war between the United States and Germany, other naval aficionados were usually quite pleased with his game.[36] "The rules alone, apart altogether from their bearing on the game, contain a mass of information . . . which cannot be found in so compact a form elsewhere, whilst . . . the strategical game will show that a number of things have to be thought of by those who command fleets in time of war."[37] To critics like Lieutenant Commander Niblack, USN, however, the game seemed too much "like one of those 'get-rich-quick' schemes."[38]

In later years, naval wargaming did achieve some popularity in the United States in the form of *Fletcher Pratt's Naval War Game*, published in 1940. Pratt employed a complicated mathematical formula that rated each ship as a function of armor thickness, speed, and the caliber and number of guns it mounted. Ships were assigned a pool of points based on this formula, and when hit by enemy fire, the pool was reduced by some number depending on the strength of the attackers. Although even Pratt himself admitted that the formula he used was "extremely arbitrary," and made a mathematical relationship out of "widely different elements among which there is no genuine mathematical relation," he countered such criticisms with the eminently practical, and potentially dangerous, argument that despite its quirks the system worked.[39] **39**

As evidence of this purported accuracy, Pratt used the example of the real-world action off the River Platte, fought in December 1939 between the German pocket battleship *Graf Spee* and the British cruisers *Exeter*, *Ajax*, and *Achilles*. Despite its heavier armor and longer-ranged guns, the German ship suffered severely at the hands of its more numerous antagonists and was forced into the neutral port of Montevideo. Rather than leaving port to engage the British once again, the *Graf Spee's* captain eventually scuttled her. Pratt argued that his rating scheme accurately showed the combined superiority of the three British ships despite the larger battleship's individual superiority. This example, overstated as it certainly is, has made Pratt's game "part of the lore of both commercial and military wargaming."[40]

Whatever the technical merits of the games of Wells, Jane, Pratt, and others, wargaming was undoubtedly making progress among both civilians and military in the first half of the twentieth century. But that progress was fitful, and dampened by reactions to the real horrors of war experienced in World Wars I and II. It was also confined almost exclusively to games with miniatures. There is little evidence of serious hobby map or board gaming. The militarily oriented board games that did exist strongly emphasized their game aspects at the expense of realism.

One early board game made its appearance in the United States during the Civil War. The game cast the player as a blockade runner, and challenged him to find a safe route through the Union squadron to the open sea. This game was little more than a maze.

Another game, which appeared in Great Britain sometime in the 1920s or 1930s, was reminiscent of the early "war chess" games, and the forerunner of the modern game of *Stratego*. Each player had an army of cardboard figures mounted on tin bases. These figures were printed on one side only, thus preventing the opposing player from seeing the ranks of the pieces. The players moved the pieces over a square-gridded map, and when one piece entered the square occupied by an opposing piece the lower-ranked unit was destroyed.

Not surprisingly, perhaps, at least two games achieved some success in Germany during the 1920s and 1930s. The first was

40 known as *Schlactenspiel,* played in a manner similar to Chinese checkers but with terrain and buildings to impede movement. Sets of the game provided for the "re-creation" of several battles from the 1813 struggle against Napoleon, and later the 1814 and 1815 campaigns. Additional set-ups dealt with the Prussian wars against the Danes, Austrians, and French, and with some of the more interesting battles of World War I.

"The historical research for *Schlactenspiel* was meticulous. For each battle, the terrain was to scale, the various troop contingents were identified on the set-up maps by unit numbers and commanders' names, a historical account of the battle was given, and each of the set-up folders included a historical narrative of the respective campaign, linking all the battles. . . . Surprisingly, *Schlactenspiel* never became very popular in militarist Germany of the 1930's. Perhaps it was too historical, analytical, academic in its approach, perhaps it just lacked good advertising."[41]

The second game was known as *Wehrschach,* roughly translated as "combat chess," and was promoted by Nazi Party organizations. It was much like chess, but was "despised by dyed-in-the-wool chess aficionados, and was too complex and abstract to appeal to minds other than chess players."[42] Similar complaints probably explain the relative slowness of board wargaming's rise to popularity in general.

STRATEGIC WARGAMING AND THE WORLD WARS

If the gaming hobby was slow to adopt and develop games played on a map or other flat board, such was not the case with the professionals. The nature, number, and quality of maps available for gaming increased rapidly in the late nineteenth and early twentieth centuries. From the original maps of imaginary or "ideal" terrain used by von Reisswitz and other pioneers, gamers moved to detailed charts of actual terrain in various scales. The availability of the different scales abetted the expansion of the domain of wargaming from its tactical roots into the higher realms of military operations and strategy.

As late nineteenth- and early twentieth-century warfare seemed to become a matter of who could mobilize and deploy to the right place fastest, the emphasis on strategic gaming to test mobilization plans increased. Because the growing size of the armies made it impractical (not to mention provocative!) to practice a full-scale mobilization, the need for some sort of simulation was apparent. The advent of strategic gaming signified a shift in **41** wargaming's focus from its origins as a training and educational device into a tool for study and research into the potential problems of mass military operations. In particular, the high-level games were adopted as important planning tools. Once again, the Germans led the way.

German Gaming

As described in the paper prepared for the U.S. Army by high-ranking German officers following World War II, the German army saw wargaming as a broadly applicable tool.[43] The term *kriegsspiel* (translated by the unenlightened as the two-word "war game," the use of which in the following quotations does not constitute an endorsement) encompassed a variety of different types of activities. "Besides serving its main purpose of assisting in the training of officers of all ranks, the war game is a means of testing new methods and checking certain combat principles."[44]

One of its earliest uses in this latter role occurred under the regime of another of the great Prussian/German chiefs of the General Staff, Alfred Graf von Schlieffen. Von Schlieffen led the German armed forces from 1892 to 1906, a period of intense war planning by all the European powers. He used wargaming and wargaming techniques, along with staff rides and standard exercises, to help him test his various plans for fighting the French again, who might this time have British and Russian aid.

Despite the failure of von Schlieffen's plan as ultimately modified and carried out by the German army under the younger von Moltke in 1914, the Germans continued to use high-level wargaming even during the actual fighting in World War I. The most well-known example is that of the final German offensive of 1918. The Germans tested and rehearsed their plan for this last-ditch attack in several strategic games, all of which indicated that it seemed to have little chance to achieve decisive success.[45]

German gaming between the wars spanned the spectrum from tactical to strategic. In addition, the Germans pioneered a new form of gaming, now known as political-military gaming. In 1929, a young officer named Manstein, who would later become one of the most brilliant of Hitler's field marshals, suggested a game scenario in which the new Polish state, recently victorious over the Bolshevik Russians, invaded German territory. Instead of being a purely military game, representatives of the Foreign

42 Ministry were invited to play the roles of the president of the League of Nations and the important political and diplomatic leaders of Germany and Poland.

The Germans used gaming for all the traditional reasons as well. New combat principles of mobility and firepower were examined in games during which one team employed the strategy and tactics estimated to be appropriate to the chosen enemy. Other games suppressed the tactical play somewhat and concentrated on exploring problems of logistics and transportation.

In the early 1930s Field Marshal von Blomberg organized several high-level games and staff studies to explore "the problems which the military and political situation had created for German national defense and, especially, to establish a theoretical basis for the joint action of the Supreme Armed Forces Command and the high commands of the Army, the Navy, and the Luftwaffe in all the important sectors of warfare."[46] At the more operational level, Generaloberst Beck employed wargaming in his 1936 effort to prepare a new manual of modern operations for the entire army. After he and his advisers had decided on the principles they thought were most important in the new conditions of warfare of their time, they called on "seasoned officers" to test those principles using wargames. "However, General Beck was completely aware of the fact, which he constantly emphasized, that all the knowledge gained in war games can never replace experience gained in actual warfare. . . . [A] war game is only one out of many aids for recognizing the demands which will be imposed by a future war."[47]

This same Beck, as Chief of the Army Staff, conducted a game only two years later, in 1938, to explore the prospects for a German invasion of Czechoslovakia. Beck used the results of the game in his attempt to persuade Hitler that the invasion of Czechoslovakia could only bring "catastrophic results . . . for Germany and all of Europe."[48] Beck lost his argument with Hitler, and his job. The Munich accord saved the Germans from finding out how prophetic Beck's game was, but led to even greater disasters in 1939.

As political tensions mounted in the period after Munich, Captain (later Grand Admiral) Karl Doenitz, chief of the German submarine force, gave serious thought to how his U-boats would have to operate in a coming war against the British. Earlier he had developed the concept of group tactics (what would be-

come known as the wolfpack) for attacks on escorted convoys. **43**
With war approaching, he also explored the tactical and operational problems of his proposed concepts by using wargaming techniques.

In the winter of 1938–39 I held a war game to examine, with special reference to operations in the open Atlantic, the whole question of group tactics—command and organization, location of enemy convoys and the massing of further U-boats for the final attack. No restrictions were placed on either side and the officer in charge of convoys had the whole of the Atlantic at his disposal and was at liberty to select the courses followed by his various convoys.

The points that emerged from this war game can be summarized as follows:

1. If, as I presumed, the enemy organized his merchantmen in escorted convoys, we should require at least three hundred operational U-boats in order successfully to wage war against his shipping. . . . Given this total, however, I believed that I could achieve a decisive success.

2. Complete control of the U-boats in the theatre of operations and the conduct of their joint operations by the Officer Commanding U-boats from his command post ashore did not seem feasible. . . . I accordingly came to the conclusion that the broad operational and tactical organization of the U-boats in their search for convoys should be directed by the Officer Commanding U-boats, but that the command of the actual operation should be delegated to a subordinate commander in a U-boat situated at some distance from the enemy and remaining as far as possible on the surface.

   I therefore insisted that a certain number of the U-boats under construction should be equipped with particularly efficient means of communication which would enable them to be used as command boats.

3. With the number of U-boats already available and with the additions which, according to prevailing construction priorities and speed of building, we could hope to receive, we should not, for the next few years, be in a position to inflict anything more than "a few pin pricks" in a war against merchant shipping.

   I incorporated the conclusions which I had reached as a result of this war game in a memorandum which I submitted to the then Admiral Commanding the Fleet, Admiral Boehm, and to the Commander-in Chief of the Navy. The former came out in strong and unequivocal support of my contentions.[49]

When war finally came, far too early for Doenitz, the concepts that the wargame had helped him to flesh-out proved all too effective for the Allied cause.

44   The coming of World War II saw the Germans making extensive use of wargames in all situations. Prior to the 1940 campaign in France and the Low Countries, and again before the 1941 invasion of the Soviet Union, games and exercises of all types were used "to prepare all the officers and noncommissioned officers theoretically for the impending operation. Each of them down to company commanders, was familiar with his preliminary duties and with the difficulties which he would have to overcome with respect to both the enemy and the terrain." As a result of such extensive preparation, "the first days of fighting went off without any friction according to the prescribed plan, and in almost no place was it necessary for higher echelons to intervene."[50]

Finally, it is fitting to end the discussion of German wargaming during the Second World War with Hofmann's original account of one of the most famous, or at least one of the most notorious, games played. It took place on 2 November 1944 in the period of sporadic American attacks on the approaches to the German Siegfried Line, which preceded the Ardennes battle (the Battle of the Bulge) of mid-December. The site of the game was the staff quarters of the Fifth Panzer Army. Let us now take up Hofmann's account.

On this occasion the staff, under the direction of Army Group Model, was supposed to rehearse the defense measures against a possible American attack against the boundary between the Fifth and Seventh Armies. The leading commanders and their General Staff officers were assembled at the headquarters. The map exercise had hardly begun when a report was received that according to all appearances a fairly strong American attack had been launched in the Hurtgen-Gemeter area. Feldmarschall Model ordered that with the exception of the commanders who were directly affected by the attack all the participants were to continue the game and use the currently received front reports as additional information for the course of the game.

During the next few hours the situation at the front—and similarly in the map exercise—became so critical that the army group reserve (116th Panzer Division) had to be placed at the disposal of the threatened army. It thus happened that the division commander, General von Waldenburg, who was present in the room and engaged in the game, received his orders one after another from the army group, the army and the commanding general in question. After only a few minutes General von Waldenburg, instead of issuing purely theoretical orders at the map table, was able to issue actual operational orders to his operations officer and his couriers. The alerted division was thereby set in movement in the shortest conceivable time. Chance had transformed a simple map exercise into stern reality.[51]

Japanese Gaming

Japan, Germany's erstwhile ally in the Second World War, probably learned about wargaming principally from the work of Meckel, who visited the islands at least twice. The Japanese War College soon introduced gaming into its curriculum, and "the successes of the Japanese Army in the Russo-Japanese War of 1904 were attributed in part to the 'lessons learned' by Japanese officers in war games."[52]

After the outbreak of European hostilities in 1939, the Japanese expanded their preparations for war and included wargaming among their research and planning tools. The Total War Research Institute was established in 1940 as part of the "powerful Planning Board which coordinated the vast, complex structure of Japan's war economy."[53] The institute's job was to explore the possible courses of action that might be open to Japan in the increasingly dangerous conditions of the time. Analytical gaming of the political-military type became one of the principal tools of the institute. Games were designed to have players represent "not only different nations such as the U.S., Britain, Russia, China, Germany, etc., but also the conflicting interests within Japan: Army, Navy, and civilian."[54]

As war with the United States began to seem more and more likely, the plans for fighting such a war became the topic of hot debate between the Naval General Staff, under Admiral Osami Nagano, and the Combined Fleet, under Admiral Isoroku Yamamoto. To analyze the effectiveness of a proposed surprise attack on Pearl Harbor, and to rehearse for its execution, a series of wargames was played in the fall of 1941 at the War College in Tokyo, including "table top maneuvers" held at the Tokyo Naval War College in mid-September.[55] Additional games were employed to aid the planning for the attacks on Malaya, Singapore, Burma, the Philippines, the Dutch East Indies, the Solomons, and the islands of the Central Pacific, including Guam and Wake.

After their brilliant initial successes, the Japanese again began to debate their future strategy. By late February 1942, the Combined Fleet proposed a plan for an operation in the Indian Ocean. "The plan was then put to the test in a war game staged on board super-battleship *Yamato*, which had just been designated flagship of Combined Fleet. The game continued over a four-day period with representatives of the Naval General Staff

46   attending."[56] The scheme played out in the game was rejected by the army, which was preoccupied with its ongoing operations in Burma and its concern about the USSR.

Combined Fleet reacted to the rejection of the western campaign by looking again to the east, but this time only as far as Midway, a little over a thousand miles from Hawaii. Midway was only the beginning of an ambitious operation the Japanese now began to test, again with wargames playing a prominent role. On the first of May 1942, the headquarters of the Combined Fleet began a series of wargames that would go on for four days. These games encompassed the entire range of operations envisioned by Combined Fleet for the next phase of the war, including the capture of Midway and the western Aleutians in early June, the seizure of key points in New Caledonia and the Fiji Islands in July, carrier strikes against Sydney and the southeast coast of Australia, and attacks against Johnston Island and perhaps even Hawaii in August.

Except for the staff of Combined Fleet Headquarters, all those taking part in the war games were amazed at this formidable program, which seemed to have been dreamed up with a great deal more imagination than regard for reality. Still more amazing, however, was the manner in which every operation from the invasion of Midway and the Aleutians down to the assault on Johnston and Hawaii was carried out in the games without the slightest difficulty. This was due in no small measure to the highhanded conduct of Rear Admiral Ugaki, the presiding officer, who frequently intervened to set aside rulings made by the umpires.

In the tabletop maneuvers, for example, a situation developed in which the Nagumo Force [the carriers] underwent a bombing attack by enemy land-based aircraft while its own planes were off attacking Midway. In accordance with the rules, Lieutenant Commander Okumiya, Carrier Division 4 staff officer who was acting as an umpire, cast dice to determine the bombing results and ruled that there had been nine enemy hits on the Japanese carriers. Both *Akagi* and *Kaga* were listed as sunk. Admiral Ugaki, however, arbitrarily reduced the number of enemy hits to only three, which resulted in *Kaga's* still being ruled sunk but *Akagi* only slightly damaged. To Okumiya's surprise, even this revised ruling was subsequently cancelled, and *Kaga* reappeared as a participant in the next part of the games covering the New Caledonia and Fiji Islands invasions. The verdicts of the umpires regarding the results of air fighting were similarly juggled, always in favor of the Japanese forces.

The value of the games also was impaired by the fact that the participating staff officers from several major operational commands, in-

cluding the Nagumo Force and the shore-based Eleventh Air Fleet, had   **47**
had little time to study the operations to be tested. The result was that
they could only play out their parts like puppets, with the staff of Combined Fleet Headquarters pulling the strings. The lack of preparation
was illustrated by an incident which occurred during the Midway invasion maneuvers. There, the somewhat reckless manner in which the
Nagumo force operated evoked criticism, and the question was raised as
to what plan the Force had in mind to meet the contingency that an
enemy carrier task force might appear on its flank while it was executing
its scheduled air attack on Midway. The reply given by the Nagumo
Force staff officer present was so vague as to suggest that there was no
such plan, and Rear Admiral Ugaki himself cautioned that greater consideration must be given to this possibility. Indeed, in the actual battle,
this was precisely what happened.[57]

Most accounts of the Japanese Midway games latch onto the
changes made to the rulings of the umpires as a prime example of
the dangers of introducing bias into wargames. Indeed, the games
were almost certainly biased; based on their fundamental premises of such grandiose operational schemes, they could hardly be
otherwise. But the point that is too often missed is that contained
in Fuchida's last paragraph. The game raised the crucial issue of
the possibility of an ambush from the north; the operators ignored the warning, a warning reiterated by the oft-maligned
Ugaki. Ugaki's change of the umpire's evaluation of the effectiveness of the U.S. land-based-bomber attack was not necessarily
blind arrogance. In the actual battle, B-17s attacked the Japanese
force on more than one occasion and failed to score a single hit!
The myth that the Japanese umpires successfully predicted the
course and outcome of the battle of Midway only to be overruled
by the overly optimistic game director is one that is in serious
need of exploding. Ignoring or changing the results of a few die
rolls did not constitute the failure of Japanese wargaming in the
case of Midway; ignoring the questions and issues raised by the
play did. The almost legendary Commander Minoru Genda, air
officer for Nagumo's staff, put his finger on the principal failure of
the game when he discussed the play of the "American" commander for the game, Captain Chiaki Matsuda. In postwar remarks, Genda stated that Matsuda's uncharacteristic American
play "might have given us the wrong impression of American
thinking."[58]

The disaster at Midway did not prevent the Japanese from
continuing their use of wargaming, but it may have made them

48   more careful in their choice of who would play the Americans in the games. After the U.S. Marine Corps's assault on Guadalcanal in August of 1942, the Japanese conducted a series of games to explore their options for destroying the U.S. forces in the South Pacific. Officers of the Naval General Staff who were intimately acquainted with the current condition of the fleet played Blue, the Japanese side.

To obtain the best possible players for the Red (United States) side, the General Staff arranged for participation by the most thoroughly informed Japanese officers with the most up-to-date contacts with the United States. They found them among some outstanding Japanese Naval Intelligence Officers who had been assigned to duty in the Japanese Embassy in Washington, and who had been interned with all Japanese nationals in the United States when war broke out. In August 1942 arrangements were completed to repatriate internees, with a mutual exchange of Embassy personnel.[59]

Before the ship carrying the returning intelligence specialists could even dock in a Japanese harbor, however, the General Staff had sent a launch to pick up the navy officers and whisk them straight to Naval Headquarters in Tokyo, "where they were held incommunicado to seal them off from all news and contacts which might affect the 'pristine information they carried concerning the U.S.'"

"The officers were told they were to play the Red (U.S.) force in war games, a task they performed thoroughly and exceedingly well. This Red team of Japanese intelligence experts demonstrated in the game that Japan's only hope was to achieve its conquests and consolidate them as early as possible, because the greater resources of the United States, once converted to full war-making capacity, would surely deprive Japan of its war potential and force her surrender."[60] Hausrath gives the source of this information as Lt. Roger Pineau, USNR, who interviewed the Japanese officers involved in the games after the war. It was Pineau's opinion that "the Japanese indulged in some sort of war game for every major operation of World War II."[61]

Russian Gaming

The Russians were somewhat slower to adopt wargaming than other European countries, but by the mid-1870s wargaming techniques were an accepted element of officer training. "By War

Department Orders No. 28 of 1875 and No. 71 of 1876 the sys- **49**
tematic instruction of officers was to be taken up by means of
written exercises and lectures on tactics under the direction of
regimental and battalion commanders. War games were to be
held in conjunction with this instruction whenever sufficient
time, rooms and other facilities were available."[62] By the early
1900s more senior officers were involved in large-scale games,
and naval war games were also held.

The results of the Russian games were not, however, univer-
sally successful. In 1903 War Department Orders No. 85 identified the causes of failure as "the inability of the directors to arouse
interest in the games; too much adherence to fixed models; a
scarcity of good directors; a lack of interest on the part of the
higher commanders; and insufficient familiarity of the participants with the tactical handling of the three arms."[63] The Russian
defeats at the hands of the Japanese in 1904–05 encouraged
greater interest in wargames on the part of the higher command,
and as the storm clouds of 1914 gathered, the Russian General
Staff played a game to test their plans for mobilization against the
Germans and the initial attacks into East Prussia.

The Russian plan envisioned an attack by two armies, one
moving to the north of the Masurian Lakes and the other to the
south. (See figure 1.) The games were played to test this plan,
and their course revealed a serious weakness in it. Because of the
separation between the two armies, forced on the Russians by the
geography of the region, the timing of the advance was crucial.
Should one army begin its attack too late, the other would be exposed to a concentrated German counterattack. The games indicated that to avoid decisive defeat the Russian Second Army
would have to begin its march three days before the advance of
Rennenkampf's First Army, *"an action not contained in the
plans. This change, so clearly indicated in the war games, was
never made in the plans or in their execution."*[64]

German games dealing with the situation identified the
same potential problem for the attacking Russians, but the Germans took the lesson to heart. The German Eighth Army under
Hindenburg and Ludendorff smashed both Russian armies, one
at a time, in what came to be known as the Battle of Tannenberg.
Unfortunately, there is precious little published documentation
of the further role of wargaming in Russian (and later Soviet)
military developments before the end of World War II.

Figure 1. The Russian/German (East Prussia) border in 1914.

British Gaming

Despite some earlier fits and starts, wargaming was seldom used by the British military until after the Franco-Prussian War. Captain Baring of the Royal Artillery introduced a set of game rules to the British Army in 1872. Baring's rules were based on those designed by von Tschischwitz, and were thus of the "later rigid" school of Kriegsspiel.

In a little over a decade, wargaming achieved some measure of acceptance in the British Army, enough to encourage the commander in chief, the Duke of Cambridge, to introduce wargaming officially into the army by an order of October 1883. After another dozen years, an official set of British Army wargame rules was published under the title "Rules for the Conduct of the War Game on a Map."[65] Perhaps the most important and influential figure in British wargaming in this latter part of the nineteenth century was a well-known military reformer named Spenser Wilkinson.

In his work *Essays on the War Game,* Wilkinson described wargames, in terms similar to those used by Sayre, as maneuvers held on a map. He cautioned: "Probably ·no form of military study is more useful if properly conducted, as certainly none is so liable to be misused." To Wilkinson, wargames were useful principally as a means to improve the tactical and strategic understanding of the participants. "The only difference from actual war is the absence of danger, of fatigue, of responsibility, and of the friction involved in maintaining discipline," minor details that just happen to be crucial in real war. "The question therefore becomes—How many men must be killed or wounded before the remainder will be induced to change their mind and go back?"[66] This emphasis on attrition and on assuming that the unmeasurable applied equally to both sides will appear again and again in different guises.

The Royal Navy also got a taste of wargaming after Captain Philip H. Colomb, RN, invented and patented a ship-to-ship game in 1878. Colomb's game, called "The Duel," simulated the detailed combat actions of two contending ships. As perhaps the first true naval wargame, it elicited a certain amount of interest on the part of both the French and Italian navies, as evidenced by the appearance of reviews in *Revue Maritime et Coloniale* and *Rivista Marittima.*[67]

As Wells's comments about the general lack of interest in army Kriegsspiel imply, wargaming never really caught on among the professional British soldiers. Nor did Colomb's game make much headway with the Royal Navy.

Wargaming, particularly the type of strategic gaming popular with von Schlieffen and his contemporaries, never had as much influence on British planning prior to World War I as it did in Germany. Perhaps this was because of the British tendency "to obey tradition and instinct, and, in the Army, an antipathy for 'professionalism.'"[68] There was, however, at least one exception to this rule, one occasion when the classic German methods were applied to study an important issue, the potential for British Army operations in a major European war.

52   Following the Crimean War of 1854–56, British foreign and military policy was deeply affected by apprehensions over an imperialistic Russia. Indeed, during the ill-fated voyage of the Russian fleet from the Baltic to the Straits of Tsushima during the Russo-Japanese War of 1904–05, the dangers of a serious conflict between Britain and Russia seemed very real. Yet, as the new century began, some perceptive officers began to see the growing Anglo-German trade rivalry and German push for naval equality or supremacy as an even more serious menace.

Initial concerns about a direct German invasion of Britain, a topic of some discussion in the German press, gave way by 1905 to concern about the possibility that war between the two countries could result from a German violation of Belgian neutrality during an attack on France. In that year, the new British General Staff decided to explore the possibility in a wargame. "The game's main purpose was supposed to be instructional, but it became the basis of British military planning for years to come."[69]

The scenario, as we would call it today, assumed that war between France and Germany had broken out in January 1905. For two months the Germans had attacked the French defenses in the same region they had struck during the Franco-Prussian war, between Sedan and Belfort. (See figure 2.) Stymied in their attack, the Germans decided to send more than 250,000 men to outflank the French defenses by marching through Belgium. Britain, the guarantor of Belgian neutrality, would thereupon be obliged to enter the war.

The game was a three-sided one. Colonel C. E. Callwell, who would probably suffer a severe case of *déjà vu* when he became the deputy director of operations in 1914, played the British commander in chief. Major General "Wully" Robertson, head of the foreign section of the Military Intelligence Department, played the German commander in chief. The Belgians were handled by Major A. Lynden-Bell, a staff officer. Several officers of the Military Operations and the Intelligence staffs assisted in game play.[70]

The game highlighted several interesting insights, one of the most distressing of which was the amount of time needed to transport any substantial British force across the Channel using the existing plans. "The most far-reaching conclusion drawn from the war game, however, was that since a German invasion

Figure 2. The French/German border in 1914.

of Belgium could be expected to succeed, France could not be expected to resist an attack on her own. This resulted, after 1906, in staff talks with the French also, and in the evolution of the Anglo-French Entente, on the strength of which France mobilized in 1914."[71]

Further information about British military gaming from before World War I and through World War II is limited. Perhaps the best-known example of their use of wargaming techniques centers around Field Marshal Bernard Law Montgomery's activities in World War II.

Montgomery, given command of the British Eighth Army facing Rommel at El Alamein, developed a typically eccentric

54   mode of operation. During particularly critical stages of his operational planning process, Montgomery would abandon his headquarters in favor of a quiet, out-of-the-way spot in which he could contemplate his options and develop his ideas without the constant interruptions of day-to-day duties. His chief of staff routinely visited the *sanctum sanctorum* to bring the commander updates on the situation and return with any new orders.

During these intense planning periods, Montgomery would test his ideas by means of wargame-like exercises with his staff and unit commanders. "Each protagonist was quizzed on specific details of capabilities, requirements, reactions to enemy moves, and what he could deliver in relation to the situation and plans before him. Montgomery required his intelligence officers to play through the forthcoming battle, with the dispositions of the enemy and his own troops spread out on a map. The staff was required to imagine themselves the enemy, to react as the enemy would, and 'to play against Montgomery [this] strange and fascinating parlor game, making move-for-move against the British.' Most of this activity was accepted practice in planning, but Montgomery played with intensity and tried always to put himself in Rommel's place. He anticipated enemy actions by close study of his adversary and always asked himself, Now what would Rommel do about it?"[72]

In some sense, British wargaming during and after World War II was limited and in many ways supplanted by a new technique, which became known as operational research. A similar effect was felt by American wargaming, which was, if anything, even less solidly implanted than its British cousin.

American Wargaming

The growth of wargaming in the United States suffered from America's traditional distrust of military professionalism and from the lack of a pressing need to develop any. After the War Between the States ended in 1865, the U.S. Army settled back into its normal routine of small posts and garrisons. The mass armies that had marched and fought over the southern states were gone, and seemed unlikely to be needed in the near future. Yet a few prophets cried out in the wilderness, realizing that America's few professional soldiers had been the heart of both the Union and Confederate high commands, and that if war with a major European power should ever threaten, the small band of professionals would once again be called upon to lead the mass of the country's **55** citizen soldiers into battle. If American soldiers were to prepare for that eventuality, they would need to supplement their limited experience of handling large bodies of troops. The new-found popularity of Kriegsspiel among the Europeans seemed an ideal tool to provide such a supplement.

W. R. Livermore is usually credited with introducing German-style wargaming to the United States. In 1879 Livermore published his two-volume work *The American Kriegsspiel*. Based largely on the work of Captain Naumann, Livermore's system was a derivative of rigid Kriegsspiel. To solve the usual problems of that approach, especially its long playing time, Livermore turned to a technical solution rather than a procedural one.

The blocks representing combat units were made of porcelain, metal, or wood. They were colored red and blue in the traditional manner, but various other colors, including golds and greens, were used to distinguish different types of combat arms and specialized units such as engineers. The playing pieces were cut to scale, so that the same blocks could be used to represent differently sized units. Depending on the scale of the map, the longest of the blocks could represent a regiment of infantry in line of battle (about 1,000 men in two ranks and elbow to elbow), or a company of skirmishers deployed with sixty-four men over about one hundred and sixty yards.

The different sides of the blocks were marked with small lines or dots called scores. The number of scores exposed indicated the fraction of a unit's strength that it had lost. Special blocks were used to record ammunition levels, fatigue, and the amount of time spent on constructing trenches or other fortifications. By these means Livermore hoped to reduce the volume of paperwork that plagued rigid Kriegsspiel.

To indicate the movement and fire of the troops, Livermore employed two different kinds of physical pointers. "Arrows" were pointed at one end and rounded at the other, "indices" were shaped like swords. Both the arrows and indices were marked with vertical lines that divided the pointers into ten equally sized segments also called scores. (Figure 3 illustrates some of Livermore's devices.)

Arrows were used to indicate the direction and volume of fire. Swords were used in a similar manner to indicate the direction and speed of march. Players could thus issue orders to their

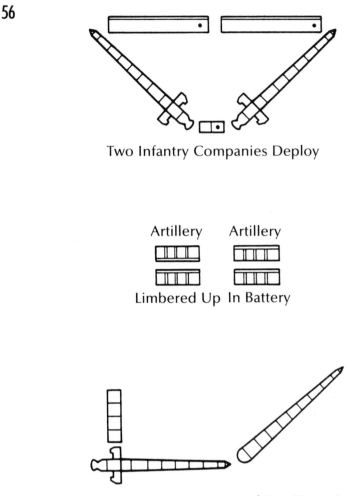

Two Infantry Companies Deploy

Artillery   Artillery

Limbered Up In Battery

Infantry Walk Two Minutes and Fire Five Minutes

Figure 3. Examples of the equipment for Livermore's game.

units simply by placing the pointers. Unfortunately, the savings in time over written orders was probably marginal at best, and the need to be careful about precisely how and where the pointers were placed almost certainly required the players to spend some amount of time adjusting their devices.

Despite Livermore's great hopes that "technology" could be the solution to the problem of speeding up play, practical experience with his game proved otherwise. The complexity and artificiality of his devices, no matter how clever they may have been, required players and especially the umpires to spend no little time mastering the conventions of their manipulation. The **57** detail and comprehensiveness of the rules and data were designed as much as possible to cover any situation likely to arise in actual combat operations. But Livermore himself realized that the devices and data did not, in fact, help speed up play to the extent he may have hoped. Instead, he encouraged umpires to rely more on their judgment and experience, guided by the vast quantities of data he provided, than on the computations alone. In the 1882 edition he wrote that "it cannot be too strongly stated that all these computations not only need not, but must not, be made in every case, after the players and umpire have had a little practice, especially if they are at all familiar with military operations. They are intended to facilitate and hasten the game and should not be so perverted as to retard it." [73]

Livermore's game was clearly of the German school, and in the opinion of some American soldiers was not appropriate to the unique conditions of the United States. One of the most outspoken of the critics was Lieutenant Charles A. L. Totten, who published his own book of wargaming techniques, *Strategos: A Series of American Games of War Based upon Military Principles*, in 1880, shortly after Livermore's first edition hit the booksellers.

Totten's purpose in writing is well described in his extended title: "And Designed for the Assistance Both of Beginners and Advanced Students in Prosecuting the Whole Study of Tactics, Grand Tactics, Strategy, Military History, and the Various Operations of War. Illustrated with Numerous Diagrams. To which is Appended a Collection of Studies upon Military Statistics as Applied to War on Field or Map." Anything else?

Because none of the foreign systems of Kriegsspiel had been available in the U.S. until his work was essentially finished, Totten was able to provide a truly unique American perspective. He avoided jumping right into the intricacies of a full-blown Kriegsspiel (as Livermore did), instead structuring his games to progress from the simple to the complex. Through his progressively more elaborate approach to the different levels of warfare in his grand tactical and battle games, Totten hoped to help the new student move up to that more advanced level, which he also covered in his advanced game.

Modern hobby gamers will find much that is familiar in Totten's games. The philosophy, physical components, and even many of the basic concepts (like the stacking of units and the area

58 of command, now better known as the zone of control) are all amazingly similar to the basic elements of modern board gaming.

Unfortunately, as the twentieth century dawned, Totten's start toward a balanced, progressive approach to the study and gaming of war seemed to be swamped in the general acclamation for military things German. As the world wars loomed, American experimentation continued with but mixed results.

In 1912, Major William Chamberlaine designed the Coast Artillery War Game for the Department of Artillery and Land Defense to train artillery officers for their possible wartime duties. Chamberlaine was somewhat unhappy with the title. McHugh quotes from the preface to the first edition: "It is considered unfortunate that some name more descriptive of its purpose could not be found but the present one has been adopted after considering several others."[74] The army's apparent uneasiness over the word game did not, however, dissuade them from cooperating with *Scientific American*, who in 1914 published an article that described gaming at the Army War College. The article noted that the technique was "used now in the instruction of every army of the world," and that wargames were not played "to see who will win, but to get results and experience, to profit by the mistakes made."[75]

Following somewhat in the tradition of Sayre, the Army General Service Schools published a book called *The Solution of Map Problems*. The stated purpose of the book was to bring about "that state of mind wherein the individual, when confronted with a situation in the field, goes about its solution with full confidence in his ability to see things in their proper relations, to weigh conditions one against the other, and to reach a sound decision without undue loss of time."[76]

Although there is some opinion that the Army War Plans Division used wargames to explore U.S. mobilization plans in the late 1930s, most American gaming seemed to remain tightly focused on training and education rather than on planning and analysis.[77] The important exception is the experience of the Naval War College, which is described in the next chapter. For the U.S. Army, wargaming meant Kriegsspiel, and Kriegsspiel meant training. The March 1941 issue of *Military Review* signified the army's attitude on the eve of American involvement in World War II. In that issue appeared a condensed translation of a

German Army booklet on wargaming written by General von **59**
Cochenhausen.[78] According to the editors of the journal, "no
better means than the 'Kriegsspiel,' or war game, has been devised for training commanders and general staff officers, approaching as it does the semblance of actual battle. It demands
definite decisions and orders for the commitment of troops, also
being conducted within the realm of time and space thereby leading to exactitude in troop leading."[79]

As was the case with the British, there is little solid evidence
concerning American use of wargaming (as distinguished from
operations research) during the course of World War II. One example is taken from the history of the U.S. Ninth Army. After
the Ardennes fighting of December 1944, the Ninth Army was
transferred from Omar Bradley's Twelfth Army Group to the
command of Montgomery's Twenty-first Army Group. As part of
the process of integrating the army into its new command, the
commanders and staffs prepared a complete and formal "estimate
of the situation." The separate corps presented their plans at a
combined meeting of all the commanders and principal army
and corps staff officers. This allowed each of the corps to understand the plans and rationales of their fellows. These plans "were
then 'war-gamed'—played out on the map—so that the action
could be thoroughly previewed and every possible contingency
discussed in detail."[80] Of course, like the similar "games" played
by Montgomery, such wargaming of operational plans in the
midst of combat was not done according to the formal systems of
Kriegsspiel popular before the war.

It seems that neither the British nor the Americans ever
quite accepted the full range of wargaming's potential value prior
to the end of World War II. Like its early days in Prussia, Kriegsspiel had a great deal of difficulty really catching on in the English-
speaking nations. The armies of both Britain and the United
States recognized some potential training value for wargames,
and even made limited use of the techniques of gaming to test
war plans. Yet the prominent role played by operational research
in the war efforts of the Allies tends to obscure what contributions
the older method may have made. The single, and stellar, exception to this assessment is the development and application of wargaming at the U.S. Naval War College. It is to Newport, then,
that our story must now turn.

2

WARGAMING AND
THE U.S. NAVAL
WAR COLLEGE

THE CREATION OF THE NAVAL WAR COLLEGE

The latter quarter of the nineteenth century was a time of social and technological upheaval for the U.S. Navy. The sudden growth in the navy's size and fighting power occasioned by the War Between the States was followed by an equally precipitous decline after that war's end. At the same time, the technological revolutions in ship construction and weapons, which the war had helped accelerate, had left the navy's core of professional officers with deep-seated feelings of uncertainty about the future and the navy's place in it.[1]

There was a feeling abroad in the country and much of the world that economic progress and industrial efficiency would soon lead to the disappearance of war and so the need for armed forces of all types. For the United States especially, insulated from the great powers of Europe by the wide expanse of the Atlantic, military security seemed virtually assured by simple geography, and military forces seemed almost superfluous. Indeed, the secretary of war, Redfield Proctor, wrote in 1890 that the "military resources of the nation have been so recently demonstrated and its network of railroads is so adapted to a rapid con-

62 centration of troops on any threatened point, that no hostile force is likely to seek an encounter with us on our own soil. A small army sent upon our shores could not hope for success; it is not probable that any large one will run the risk."[2]

If civilians believed the military to be unimportant, naval officers felt themselves unappreciated. The ships, what remained of them, were old. The officer corps was plagued by the twin specters of a rigid system of seniority and an overabundance of seniors. It is little wonder that officers with ambition and imagination found the naval service less than satisfying. "Confronted with the combination of public indifference and professional stagnation, many officers became mere timeservers while others buried themselves with great thoroughness in some technical branch of their calling, ordnance or electricity, engineering, or international law."[3]

Determined to pull the navy from the pit of its despair, a small band of "reformers" took the intiative to turn the navy into a true profession. Led by Commodore Stephen B. Luce, the reformers sought no less than the complete transformation of the navy's officers, not into mere technicians, but into well-educated, well-rounded masters of the tools and techniques of a unique naval art. In addition, they sought to secure a place for these new naval professionals in the public's appreciation and understanding by demonstrating the importance of that profession for the general life and well-being of the nation and its people.

It was a tall order, but one Luce accepted as a challenge worth his commitment. In 1881 Luce was appointed commander of the newly formed Training Squadron in the Atlantic, with his flagship, the USS *New Hampshire*, anchored off Coasters Harbor Island only two miles north of the city of Newport, Rhode Island. The ship's executive officer was a young lieutenant named William McCarty Little, whom Luce had known as a student at the Naval Academy while Luce had been on the staff there. Operating from this base in Narragansett Bay, Luce embarked on a concerted campaign to establish a higher school for the professional education of naval officers.

In October 1884, after extensive efforts both within and outside the navy, Luce succeeded in getting the secretary of the navy, William Chandler, to sign General Order 325, formally establishing the Naval War College on Coasters Harbor Island. A year later the college became a reality, with Luce, detached from his command of the North Atlantic Squadron, as its first president. **63** Among his staff were a Captain Alfred Thayer Mahan, and the now retired Lieutenant William McCarty Little.[4]

McCARTY LITTLE AND THE BEGINNINGS OF NEWPORT WARGAMING

McCarty Little, as he was known to his friends, began his navy career in Newport in 1863, where he entered the Naval Academy during its brief hiatus there during the Civil War. (Annapolis was a bit too close to the front lines.) Graduating in 1866, McCarty Little served in various sea and shore assignments, many of which enabled him to maintain his association with the city he had grown to love. In 1876, however, he suffered an accident ashore, which cost him the sight of one of his eyes. He was able to return to active duty and served well for several more years until one of those unfortunate and inexplicable bureaucratic actions resulted in his discharge for medical reasons in 1884, after twenty years in the navy. Inevitably, he chose Newport as his new, and final, home port.

Just as inevitably, it seems, he became deeply involved with the activities of his old friend Luce. Although his role in the establishment and the organization of the Naval War College during its first year of existence is not fully recorded, "McCarty Little assisted in the preparation of teaching aids, established and maintained a library, and fulfilled administrative duties of a routine nature."[5] His responsibilities expanded the following year.

Luce had been detached to resume command of the North Atlantic Squadron, and Mahan had assumed the role of War College president. Plagued by bureaucratic and political problems, Mahan leaned heavily on McCarty Little for both moral and practical support. For McCarty Little, an unofficial member of the staff, that support included the preparation of maps and charts for Mahan's lectures on strategy and tactics, and also the preparation and delivery of a lecture of his own. The latter, simply titled "Colomb's War Game," was the first on the subject to be given at the college. It was not the last.

Little's budding interest in wargaming was almost certainly stimulated by his association with army Major W. R. Livermore, author of *The American Kriegsspiel*, who was stationed across the harbor from the Naval War College in Fort Adams. The two men worked together on a unique project conceived by Rear Admiral

64 Luce, who had left the college in body to take command of the North Atlantic Squadron, but had not strayed far in spirit.

During the summer and fall of 1887, Luce brought the ships of his North Atlantic Squadron into Narragansett Bay for an extended period of time. He took advantage of his position to place them at the disposal of the college, in keeping with his notion that the education of naval officers required both theoretical and practical instruction. The end result was a joint army-navy maneuver, which Livermore helped the War College plan.

Two maneuvers were held in the bay during October and November. The first simulated a night torpedo attack on the ships of the North Atlantic Squadron. The second was an elaborate attack on the city of Newport itself. For the latter, the squadron ran through a minefield, past Fort Adams at the entrance of the bay, and suffered a torpedo attack as it passed the Navy Torpedo Station on Goat Island. It then carried out an amphibious landing at Coddington Point to the north of the War College's home on Coasters Harbor Island. Livermore served as the umpire for the maneuver, with McCarty Little acting as his assistant. At the conclusion of the exercise, a critique was prepared and discussed at the college.[6]

Unfortunately, the new secretary of the navy, William C. Whitney, was not enamored of Luce, his maneuvers, or the Naval War College. The experiment would not be repeated until after the turn of the century. In the midst of a raging bureaucratic war over the future of the college, McCarty Little's suggestion to use steam launches in place of actual warships for the college's "school of application" ran into difficulties over acquiring even these poor substitutes. As a result, the college came to rely more and more on wargaming as a supplement to its curriculum of readings and lectures.

McCarty Little, now an official member of the War College's staff, expanded his single lecture of the previous year into a series of six presentations for 1887 and conducted his first actual wargame at the College.[7] Little continued his lectures in 1888 and 1889, and he also continued refining and developing his own ideas. Livermore had become even more closely associated with the college during this period, and in 1889 he lectured there on military strategy and tactics. But the political travails of the college continued.

From 1891 to 1892 courses were suspended during con- **65** struction of a new War College building on Coasters Harbor Island, but McCarty Little did not suspend his work on wargaming. While continuing to refine his own techniques, he also embarked on a course of translating important foreign works dealing with naval strategy and tactics. Not surprisingly, his first translation was one dealing with wargaming, the Italian rules published in 1891 by Lieutenant A. Colombo of the Italian Navy.[8]

McCarty Little followed a two-year absence from the college (to serve as the United States's special commissioner to Spain for the 1892 Colombian Exposition) with a rapid reentry into the swing of things at Newport. In addition to the incessant political battles waged against opponents of the college, Little returned to his work on wargaming. A single game had been conducted during 1892 while he was away, but it was not completed. During the following year, classes were not held at the college, but the staff continued to work on refining gaming techniques. In 1894, under newly appointed President Captain Henry Taylor, gaming became an integral and permanent part of the course of study for all students.

Taylor, possibly with the encouragement and certainly the assistance of McCarty Little, integrated wargaming with that part of the students' course in which the class was assigned a problem. Typically in those early years the student problems focused on coastal defense, and each student was required to develop a complete solution to the problem. These solutions were then examined, tested, and critiqued in a wargame.

The wargames, which had had such a fitful birth at Newport, suddenly exploded in their popularity and prominence. The news media reported that "the War College has taken a new and successful departure, and the year's work just closed has been peculiarly practical and progressive. It consisted, first and foremost in working out a problem in strategy—an application to American naval tactics of the 'Kriegsspiel' to which the German Army, and particularly the officers of the General Staff, owe their high efficiency in mobilization and strategic movement. . . . That complete preparedness against all probable contingencies is the ultimate aim of this institution; and in the absence of an American General Staff, naval officers are here to determine beforehand what an enemy must or would be likely to do in attack-

66 ing us by sea, and what, under each set of circumstances, is the best way to repel him."[9] The secretary of the navy himself, Hilary A. Herbert, visited the War College in 1895 and spent his entire sojourn there observing the play of the games. He was reportedly very "well pleased by what he saw."[10] In an 1897 letter, the new assistant secretary of the navy, a Mr. Theodore Roosevelt, wrote, "When I come on to Newport, I want to time my visit so as to see one of your big strategic games."[11]

The strategic game referred to in Roosevelt's letter was only one of the three types of games whose rules McCarty Little compiled for use at the college. The simplest game was the Duel, or Single-Ship game. The Fleet Tactical game dealt with the maneuvers and engagement of two competing forces already in contact with each other. Finally, the Strategic Game examined the wider dispositions and operations of fleets involved in a major campaign.

In the Duel, ships were represented by cardboard or celluloid cutouts roughly three inches long. In the earliest days, the playing surface was a piece of paper marked off into grid squares to assist in the movement of the ships and to help record the action for later analysis. A similar grid, marked off on a scale of ten inches to the mile (one of the popular scales for land wargames of the period), was used for the Fleet Tactical game. A standard navigation chart served as the game board for the strategic game. As the size and scale of the games increased, the paper game boards became boards in a very literal sense. The grid was painted on a wooden board and laid across low sawhorses. Even later, after World War I, these boards also proved too small, and play was transferred to the deck (or floor) of a large room. The scale was reduced to four inches per thousand yards.[12]

In the duel and tactical games, dice were used along with charts of hit and damage probabilities to determine the outcome of gunfire and torpedo attacks. The strategic game adopted a more aggregated method of resolution: "If two fleets meet with odds of 2 to 1, the inferior will be removed; with odds of 3 to 2, the inferior loses one-half; with odds of 4 to 3 the inferior is destroyed but the superior is crippled for the remainder of the game."[13]

The duel game was never very popular at the War College, probably because of the limited scope of decision making that it provided. It was discontinued in 1905. The strategic and tactical games, however, remained mainstays of Newport gaming until after the Second World War. Those two types of games could be played separately, or they could be incorporated as part of the same exercise.[14] In the latter case, the staff of the college would prepare a hypothetical scenario and divide the students into two groups. One group would represent the United States, or Blue Navy; the other group would play the role of the opposition. Both teams then prepared plans and issued orders under the rules for a chart game. Once the opposing fleets closed within range of each other, the chart game ended and play shifted to the game board for resolution at the tactical level. Once the game was finished, the War College staff would analyze the performance and plans of the players and discuss with the students the strengths and weaknesses of the opposing strategies and tactics.[15]

The analyses of the games, together with the other work done by the students and staff of the college, served as the basis for many important insights, which were later incorporated into the navy's planning and doctrine.[16] Some examples include the fact that "in 1896 the College recommended that oil fuel be explored as a substitute for coal, and in 1901 concluded that the all-big-gun warship had substantial advantages over the mixed caliber type then being used by all major navies; both conclusions were well ahead of their time."[17]

In one of his last major articles, McCarty Little pointed out one of the principal early contributions of the techniques he had so staunchly pioneered. "The principle of concentration of the fleet . . . was the direct result of a strategic game. . . . Dissemination had been our rule for years, i.e., the ships were divided more or less impartially among the stations 'to show the flag' as the expression was; and at that time the same rule was general with other nations. At the beginning of the game most of the conference [class] had never entertained a suspicion that the custom was not perfectly correct; but at the end there was but one voice, and that strong and outspoken for concentration. But this view, which required but the time of one game thoroughly to capture the entire conference [class] took many a weary month before by mere argument it could convince all of those of our naval authorities who had not had the privilege or opportunity of 'seeing with their eyes.' It was some time after this that England adopted the same principle."[18]

68   In the introduction to this same article, however, Little revealed his uneasiness with the term "war game," that same uneasiness that plagues professional wargamers even to this day:

> In embarking on this lecture I would like to say, by way of preface, that the name Game, War Game, has had much the same depreciating effect as the term Sham Fight has had with regard to field maneuvers. To avoid this the Army has had recourse to the expression Map Maneuver. We, of the Navy, may in like manner say Chart Maneuver, and we have lately decided so to do. There is a further reason why it is well for us to prefer that term, namely, that it accentuates the fact that the strategist's real field of operations is the chart, just as the architect's real field is the drawing board; indeed, Jomini calls Strategy "War on the Map."
>
> Still, war itself has been declared to be a game, and rightly so, for it has the game characteristic of the presence of an antagonist. It has, however, another characteristic which differentiates it from most other games. The latter are played for sport; and good sport requires reasonable chances of winning for each side, and aims to give amusement even to the losers. In the game of war, on the other hand, the stake is life itself, nay, infinitely greater, it may be the life of the nation, it certainly is its honor.[19]

Little's perception of the similarities and differences of war and games deeply influenced his work at Newport. A critical element in the play of the War College games, and one that distinguished them from Clerk's early efforts and the contemporary map maneuvers and tactical rides championed by Sayre, was what Little himself described as "the existence of the enemy, a live, vigorous enemy in the next room waiting feverishly to take advantage of any of our mistakes, ever ready to puncture any visionary scheme, to haul us down to earth."[20] This opportunity to pit one intellect and will against another was seen as an essential element in the education of a naval officer, and some of the War College's staff came to see the wargames as more valuable than fleet maneuvers.[21]

The wargame came to be the college's primary means of tying together the broad principles that served as the basis for its educational program under Luce and Mahan to the more practical requirements of command stressed by the likes of Sims and Fiske after World War I. McCarty Little worked prodigiously to help the Naval War College manage this period of transition in the early 1900s. He was instrumental in helping to persuade the navy to adopt the "military planning process," then known as the "applicatory system." This system "was a method of teaching based on the idea that military principles were learned best by application. It consisted of three parts: the estimate of the situation, the writing of orders, and the evaluation through gaming or maneuver board exercises. Its purpose was to permit officers in command situations to exercise intelligent options for the resolution of problems rather than to be slavishly bound to a method conceived at a higher level."[22] McCarty Little argued that "it is the war game that had led to the adoption of the system," and that it was "the game [that] sought the method and not the method that sought the game."[23]

Yet McCarty Little and the Naval War College, though convinced of the value of wargaming, were only too well aware of its limitations. During his tenure as president in 1897, Captain Goodrich addressed the following words to the graduating class: "I am confident . . . that you have derived much benefit from the tactical games, which have at least taught you some things which a fleet should not do. . . . The single-ship game has made a distinct step forward through the introduction of the torpedo as a weapon. Experience and study will improve this as well as the other games, so that they may more nearly represent the conditions of actual warfare. It should be borne in mind, however, that a reasonable approximation is the best we can hope for . . . because of the imperfections that must necessarily exist in this mimic warfare, its results can not be accepted in their entirety, but must be analyzed and digested before they can be made the basis of future campaigns."[24]

When William McCarty Little died on 12 March 1915, the Naval War College had grown from its troubled beginnings into a powerful force for the development of naval professionalism in the United States. The tool he had helped to mold and develop from a mere curiosity into a central element of the War College experience was firmly entrenched in the navy's future. Francis McHugh, one of his true descendants in his dedication to both gaming and the Naval War College, summarizes McCarty Little's role in Newport gaming:

The inclusion of Little's lectures in the college curriculum represented the first official recognition of war gaming in this country, and very likely its first official recognition by any navy. The scheduling of such a series at a time when the majority of naval officers had little patience with theoretical pursuits, and when the very existence of the college was in doubt, required an unusual amount of courage and foresight

70 on the part of the then relatively unknown President of the College, Captain Alfred T. Mahan, U.S. Navy. . . . Except for one short period, he [Little] remained at the College until his death in 1915. Congress, by a special act, appointed him to the rank of captain in 1903.

Throughout his long distinguished War College career, Captain Little was concerned with war gaming. His knowledge, experience, and enthusiasm made possible the orderly development of gaming at the College; his continuous service and prestige insured such development. He seems to have been the world's first professional war gamer.[25]

THE FLAWED ORACLE— INTERWAR GAMES AND THEIR RELATIONSHIP TO WORLD WAR II

Little's death heralded the end of the beginning for Naval War College wargaming. During his long career Little had seen the War College place a great deal of emphasis on the use of games as an analytical tool, applying them in the efforts to help institute a solid professional system of planning. For these purposes, he could afford to aggregate the results of combat actions to a high degree without unduly affecting the utility of the results. In most of the games of the period, all ships of a given type were assumed to have the same characteristics for the purposes of damage assessment.

This simple technique proved inadequate to meet the requirements imposed on gaming after the First World War. During the battles of that war the capabilities and weaknesses of different classes within a particular ship type (such as the tendency of British battle cruisers to blow up in the Battle of Jutland) proved to be very important. At the same time, the increasing prominence of the submarine and aircraft in naval warfare dictated the need for updated methods.

When Admiral William S. Sims assumed the presidency of the Naval War College (for the second time) in December 1918, his focus was on educating naval officers in the practical elements of command at sea. Instead of concentrating on the theoretical or on the material and technical, Sims stressed his belief that "the art of command and coordinated effort, should be given precedence over all other consideration."[26] The college's assumption of a principal role in navy planning, which had occurred in the ten or so years prior to the entry of the U.S. in the World War, was to be balanced by renewed emphasis on education.

Thus, Newport wargaming came to focus on educational games, "that is, games conducted for the primary purpose of providing the players with decision-making experience."[27] The strategic games were improved by increasing the numbers of forces involved and by adding aircraft and submarines. The tactical games followed suit, and their play and analysis contributed substantially to the development of ideas about how to employ the aircraft carrier. Wargaming, though changed in emphasis, continued to play an important part in the growth of the U.S. Navy's professional spirit and competence.

In a 1923 article, Admiral Sims described the place of wargaming in the navy.

The principles of the war game constitute the backbone of our profession. . . . At the Naval War College our entire fleet with all of its auxiliaries, cruisers, destroyers, submarines, airplanes, troop transports, and supply vessels, can be maneuvered on the game board week after week throughout the college year against a similar fleet representing a possible enemy—all operations being governed by rules, based upon the experience of practical fleet officers, and upon the immutable principles of strategy and tactics that the students are required to learn. There is no other service in the career of a naval officer that can possibly afford this essential training. In no other way can this training be had except by assembling about a game board a large body of experienced officers divided into two groups and "fighting" two great modern fleets against each other—not once, or a few times, but continually until the application of the correct principles becomes as rapid and as automatic as the plays of an expert football team.[28]

In 1922, as an outgrowth of Sims's emphasis on realistic education, the Naval War College adopted a new and much more complex system of battle-damage assessment. This system "was based on actual armaments and actual ships. Known as the 'War College Fire Effect System,' it was designed to lead 'to sound tactical conclusions when used in connection with game board problems,' and to furnish 'a means of making substantially accurate relative strength comparisons of ships and forces.'"[29] This new system, developed, extended, and improved over the years, served as the basis for tactical play throughout the interwar period. It also reflected a subtle and not altogether positive divergence in the philosophies underlying the strategic and tactical games.

Strategic Gaming Between the Wars

Unlike the tactical games, the form of Newport strategic games changed little in the interwar period. The basic tool was the

72 chart, on which the forces were represented by symbols. As before, the staff presented the strategic scenario to the opposing players who then prepared their assessments of the situation and their plans of campaign. The players submitted their plans to the game director and the latter determined an appropriate length of time that the first "move" would encompass. The player teams retired to separate rooms where they plotted their maneuvers on overlays, which were used by the control group to update a master plot of both side's activities. The game director assessed the new situation, gave each side an intelligence update (with allowance for the fog of war), and defined the duration of the next move. Once the principal opposing forces closed to engagement range, the director either terminated the game or transferred play to the tactical level. The only change made to this classic approach during the interwar period was the addition of an intercom system and pneumatic tubes to speed the flow of moves and information.[30]

When compared to earlier strategic gaming, however, the interwar games displayed a somewhat different character, despite the similarity of techniques. For one thing, the scale of forces was much greater and now included aircraft and submarines in some numbers. Instead of being concentrated on campaigns resulting from a foreign threat to the U.S. coast, the games also explored the broad expanses of the Pacific Ocean and a possible American counteroffensive thrusting from the eastern Pacific to the Philippines. In addition, the games were more closely integrated into the course work of the strategy department. Although the staff encouraged free discussion of situations and options during the course of play, they also prepared a "staff solution," which served as the basis for postgame discussion and critique of student plans and actions.[31]

The archives of the Naval War College have preserved the records of more than 300 wargames played during the interwar period. Of these, more than 130 emphasized campaign-level or strategic play, with all but nine of those focused on a possible war with Japan. During the course of these games, the discussions and critiques of players and staff evidenced a growing appreciation for the strategic realities that would later surface during the war in the Pacific. From an initial focus on a Mahanian vision of an early, decisive clash of the battle fleets, which would decide

the outcome of the war in an afternoon, the games evolved into a **73**
more realistic and grimmer vision of a prolonged struggle, not
just between fleets, but between nations and societies. During the
process, the young officers who would rise to the command of
task forces and fleets in the coming war saw their perceptions of
reality torn apart and then rebuilt.[32]

Much has been made of the famous statement of Admiral
Nimitz that nothing that happened in the war was a surprise except the kamikazes.[33] More interesting, and probably more in-
dicative of the value of the games in the development of Nimitz's
and the navy's strategic thinking, are the words he wrote in his
1923 thesis: "To bring such a war to a successful conclusion Blue
must either destroy Orange military and naval forces or effect a
complete isolation of Orange country by cutting all communications with the outside world. It is quite possible that Orange resistance will cease when isolation is complete and before steps to
reduce military strength on Orange soil are necessary. In either
case the operations imposed upon Blue will require the Blue
Fleet to advance westward with an enormous train, in order to be
prepared to seize and establish bases en route. . . . The posses-
sion by Orange of numerous bases in the western Pacific will give
her fleet a maximum of mobility while the lack of such bases imposes upon Blue the necessity of refueling en route at sea, or of
seizing a base from Orange for this purpose, in order to maintain
even a limited degree of mobility."[34]

Over the course of the interwar strategic games the type of
careful and perceptive thought exhibited by Nimitz reshaped the
way the navy came to think of its role in a future war with Japan.
One historian of the War College games has argued that "the repeated strategic gaming of Orange war forced the Navy to divest
itself of several former 'reality-assumptions':

- The notion that war at sea was defined according to a formal,
  climactic clash of battle fleets, and that naval strategy consisted
  of maneuvering one's fleet to bring the adversary to decisive
  engagement.
- The belief that superior peacetime naval order of battle was
  equivalent to available force in war, that a peacetime treaty
  status quo would persist indefinitely, and that only traditional
  naval weapons according to traditional hierarchies of importance would be necessary to defeat the enemy.

74
- The assumption that naval war across an oceanic theater could be conducted quickly, and that enemy advantage in strategic geography was marginal both to strategic planning and to the conduct of naval operations in war.
- The hypothesis that war with Japan would be limited in forces engaged, in objective, in belligerent participants, and in time."[35]

By emphasizing the broad questions of strategy and managing a fleet at the expense of downplaying some of the technical and tactical details of combat, the War College's strategic games left the players free to explore their options and to teach themselves the deeper truths of the war that was coming. Based on the insights afforded by strategic gaming, the navy began to explore the requirements for a measured, step-wise offensive campaign to span the Pacific, requirements not just for the navy, but for the entire national political and military apparatus. "Gaming reality forced the Navy to seize a set of strategic concepts about the conduct of future war which had the capacity to redefine the very nature of America's role in the world."[36]

Tactical Gaming Between the Wars

Contrast this assessment of strategic gaming with the evaluation of the contemporary tactical gaming made by a student of the latter games. "While the war games played by the American Navy prior to World War II hinted at tactical problems, alternative hypotheses were rarely tried and apparently not desired. The games were a poor source of analysis made even poorer by a failure to seek alternate solutions for the problems discovered and a sense among the American naval hierarchy that the results of the game were adequate for the purpose of tactical evaluation. Missing from the study of game results was the realization that the analysis of the outcomes is the beginning of wisdom, not the end product."[37] Why this dichotomy?

When Harris Laning became the president of the Naval War College in the early 1930s, he drew on his experience as a student at the college and as the head of its tactics department to increase the emphasis on the tactics of combat action between battle fleets. He created a department for statistical studies to compile and summarize the available data on weapon effectiveness and the operations of different types of ships as well as aircraft and submarines. Furthermore, the department was to "keep full records of the details of all games played and from study and analysis of these records to ascertain the salient points and features relating to gunfire, torpedo fire, bombing, smoke screens, damage received and inflicted by the different types, use of aircraft, etc., together with statistical data as to material features."[38]

The end result of such fascination with purportedly hard data was an extremely detailed system of damage assessment that rated ships in terms of their ability to sustain damage from fourteen-inch shells. Each weapon was rated in such terms for its ability to damage each target. "The losses were computed or extrapolated from the results of target practice and armor penetration studies, and such factors as rates of fire and angles of impact of hits were carefully considered."[39]

The tactical game was played in time-stepped moves of three-minutes' duration. During each move there were discrete phases for movement, search, and communications, all separate from the allocation and assessment of the effects of fire. Screens were placed on the game floor to limit the information available to the players to that which they could themselves see or which were communicated to them by subordinates. "Visibility, sea state, wind, and the relative position of the sun were all considered in the evaluation of gunfire between ships. . . . The game rules for maneuvering forces were extensive, and integrated with the restrictions on visibility and communications . . ."[40]

Despite all their detail and apparent realism, however, the complex rules of the tactical game failed to account for one of the most crucial elements of the reality of ship-to-ship action during the Pacific war. In the limited number of surface combats fought by the U.S. Fleet, virtually none of which followed the expected, classic battle-line versus battle-line scenario, sudden engagement, rapid decisions, and catastrophic losses were the order of the day. The leisurely pace of the three-minute move was nowhere to be seen; "Three minutes was a long time in a night battle at point blank range; information about opposing forces was scant and often confusing. The game at Newport, which stressed thorough planning and the measured engagement of the enemy in gradual attrition at long range, was poor preparation for

76 the mayhem of the close-in clashes of the Solomons where forces rapidly approached each other at point blank ranges, and ships' combat lives were measured in minutes."[41]

The same commentator softens his blows somewhat by declaring that "the blame cannot be laid entirely upon the game and its rules; they merely reflected the predilections of the American Navy at the time and sought to frame these inclinations into tangible form for reduplication on the game floor. . . . Gaming's failure in the case of the Solomons was more the failure of tactical conceptualization than of the gaming system."[42] Yet, as we saw earlier, that same navy was fully capable of discarding its preconceived strategic notions, something it failed to do with its tactical doctrine.

The failure of interwar tactical gaming can be traced to a failure of perspective. For the first time in its history, Naval War College gaming came to be dominated by the tool, particularly the Fire Effects System, to the detriment of the player. The almost overwhelming pseudo-realism of the charts and tables seemed to reinforce the validity of the doctrinal assumptions of the dominance of the gun over the torpedo and the centrality of the battle-line engagement. Immersed in the intricacies of maneuvering their ships and firing their guns, the players lost sight of the larger questions. Instead of inculcating the players with the habits of quick thinking and reaction to the unexpected, which would have served them well in the Solomons and other actions of the Pacific war, the interwar games lulled them into a false sense of comfortably paced, discretely phased combat. The result was a hard and bloody reeducation in the waters of the south Pacific.

EARLY POSTWAR GAMING

As the storm clouds of World War II gathered on the horizon, the Naval War College underwent a series of changes. The president of the college, as senior officer in the area, was given added responsibilities as commandant of Naval Operating Base, Newport. The courses of the college were restructured and shortened considerably, leaving little time for wargaming, although gaming facilities and equipment were sometimes used to restage actual battles from the war. In addition, the college became one of the sites of the Army-Navy Staff College, a 1943 creation designed to

acquaint officers from each service with the characteristics and **77**
practices of the other in hopes of facilitating joint operations.[43]

After the war, as the navy and the college struggled to adapt
to a new environment, the number of games was reduced at the
same time that their scope and level increased. Yet gaming was
still considered an important element of the War College experience, and plans to modernize the gaming facilities were considered as early as 1945.[44] Meanwhile, the game "rules and the
damage assessment system which had been employed for all College games were replaced gradually by less-rigid rules tailored to
the purpose, the level, and the scope of each separate game."[45]
By the early 1950s, each wargame was designed with its own set
of rules and damage-assessment criteria. One of the earliest uses
of these individualized techniques was made in the new logistics
course under the supervision of Captain (later Rear Admiral)
Henry E. Eccles. The students in this course, established in
1947, "solved naval problems, working them out on the game
board. These ranged from a quick tactical problem to a major
one involving a global war."[46] The junior and senior classes in
the Command and Staff course also continued the tradition of
solving game problems, with the members of the junior class
usually taking on subordinate roles for each side.

During the first twelve years after the war, War College
gaming focused strongly on education and emphasized operations at the level of task groups and higher. During the later years
of the period, the techniques for playing the board game changed
dramatically in an effort to speed play. The players operated from
separate rooms, sending their moves to the control group in
much the same manner as had been employed for the old chart
game. "The control group made the moves on the game board,
evaluated interactions with the aid of simplified procedures, and
relayed CIC-type intelligence to the players."[47]

The strategic games changed as well. They no longer concentrated on strictly naval campaigns, but also included joint and
combined operations. National-level strategic games were introduced, incorporating political and economic factors.[48] Moves in
such games "frequently involved days rather than hours, and
evaluations were based largely upon the professional judgment of
the umpires. During the 1956 and 1957 national-level strategic
games, leased teletype lines connected the College and The

78 George Washington University, and the high-speed computer of the latter's Logistic Research Project was employed to check the logistic feasibility of opposing plans and to assist in the assessment of damage."[49]

NEWS AND THE ARRIVAL OF ELECTRONIC WARGAMING

The link to the George Washington computer was the herald of the future of Newport wargaming. The modernization plans laid in the late 1940s came to fruition in 1958 when the Navy Electronic Warfare Simulator, the NEWS, was commissioned. The NEWS occupied the three floors of the central wing of the War College's Sims Hall, and had taken thirteen years and $7.25 million to complete.[50] In a very real sense, it was "the post–World War II successor to game board and chart maneuvers introduced by William McCarty Little in the 1880s."[51] Not only did it make the old boards and ship models obsolete, but it also got the attention of the fleet commanders, who saw its potential utility as an operational-training device.

The name of the NEWS can sometimes cause confusion today. NEWS was not a simulator of electronic warfare, but was rather an electronic simulator of warfare. The system was built around a large-screen display that dominated the bottom floor of the facility and served as the principal tool by which the umpires kept track of the action. The second floor contained shops and air-conditioning equipment, but also had a balcony overlooking the "game floor" and a small conference room. The third floor was the player area, composed of twenty command centers, ten each for "Green" and "White," the original, non-traditional colors used to denote the opposing sides. (See figure 4.) Command centers were equipped with displays, control facilities, and also voice-, digital-, and hard-copy communications facilities.[52]

The NEWS was made up of three primary subsystems: the maneuver and display, damage computer, and communications subsystems.

The maneuver and display subsystem had the capability of displaying the location and movement of up to forty-eight separate "forces" (and also fourteen fixed forces or targets) on both the master plot and the displays in the various command centers. A force could be used to represent anything from a single ship or

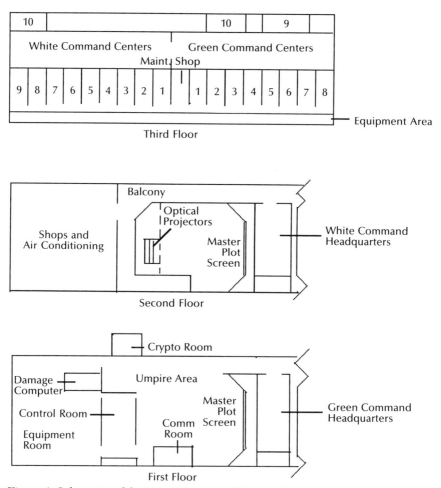

Figure 4. Schematic of the Navy Electronic Warfare Simulator (NEWS) facility. (Adapted from McHugh, Francis J. *Fundamentals of War Gaming*. 3rd ed. Newport, RI: U.S. Naval War College, 1966)

aircraft to an entire task force. Forces were projected on the master plot screen, with the shape and color of the image reflecting the type of force and the side to which it belonged. Command centers received less information, their screens displaying all forces as a standard-sized "blip," which in itself carried no information about the force it represented. The maneuver and display system integrated the movement orders made in those command

80   centers and presented to the players the type of information they would receive in a Combat Information Center (such as azimuth, course, speed, and identification of a contact).

The damage computer was a specially designed, time-shared, analog machine. Its database was capable of storing the effectiveness characteristics of twenty different weapon types against twenty different target types. Each weapon-target combination was characterized by the maximum engagement range, the probability of hit at zero range and at maximum range, the change in hit probability as a function of range, and the incremental damage on the target per hit. Each force was identified according to its target type and could employ up to four weapons. If a force commander elected to engage a target, he would press weapon controls in the command center that activated the damage computer. The results of the attack were reported to the umpires and the command centers in terms of the fraction of effectiveness the target had remaining. Lost effectiveness translated into a reduction of weapons capability and possibly of maximum speed. There was also provision to allow damage assessment to "be manually modified by umpire decision due to operational considerations, resulting in changes in hit probability and incremental damage per hit." [53]

The third major component of the NEWS was the communications subsystem. This system integrated "eight simulated voice radio networks, a teletype, and an intercom facility in each command center." [54] An additional system allowed the transmission of digitally coded information in a one-way link from command centers to umpires. Like the manual games of the early 1950s, players could either control their forces directly, using the equipment and procedures available in the command centers, or they could employ the communications facilities to pass maneuver and fire orders to the umpires, who would then carry out those orders using the NEWS systems and report back to the players. This indirect mode of play was especially appropriate when flag officers were involved in the game. [55] Over time it became the more popular technique for using NEWS, limiting the amount of specialized knowledge and equipment manipulation required of the players.

NEWS was designed to play two-sided games in continuous time. The speed of the clock could be set to real time or to twice or four times real time. Detections occurred automatically, as did the calculation of engagement outcomes. The umpires, relieved of these "tedious and time-consuming details . . . [were] free to watch the larger picture, to inject their professional judgment into the game."[56] This automatic aspect of play was considered of prime importance because NEWS was designed to be an "educational rather than analytical war gaming device, . . . [presenting] Naval War College and fleet officers the opportunity to gain significant combat command experience in a realistic setting and under the press of real time."[57]

Unfortunately, from the day it first came on line NEWS suffered from several limitations. It had originally been conceived of as little more than a fancy modern update of the game board (one of its preliminary names had been the Electronic Maneuver Board System), and this ancestry showed all too clearly.[58] As originally designed, the system could represent only four different sizes of areas of operation: 40 by 40, 400 by 400, 1,000 by 1,000, or 4,000 by 4,000 nautical miles. Force positions were restricted to a fixed 4,000 by 4,000 grid, independent of the size of the area (that is, 1,000 grid spaces represented 1,000 nautical miles at the largest scale and 10 nautical miles at the smallest), and although the size of the playing area could be altered during a game, grid positions were not automatically adjusted to reflect the dimensions of the new playing area. Each side was restricted to a maximum of twenty-four forces, and the system did not automatically track or record the positions or status of those forces over time. Furthermore, although the time scale was flexible, changes in the game clock rate affected only the horizontal component of a force's movement; changes in altitude and depth continued to require a fixed amount of real time. Finally, damage probabilities were determined only as a function of range, and target azimuth had no effect. Weapon time of flight was ignored, as were the effects of a near miss (weapons had to hit to score any damage, limiting the ability to reflect the use of nuclear weapons).[59]

To get around these limitations, a whole host of "scaling factors and other gimmicks" were developed. Scaling factors allowed the approximate representation of areas of different sizes by adjusting the speed controls of the forces and the clock rate. Other techniques were employed to adapt the damage computer to simulate fuel expenditures, and the weapons characteristics to

82 simulate fighters. In addition, manual gaming techniques were used to supplement the NEWS, especially for play at higher levels of command.[60] Such tricks of the trade could only be created and employed by the experts at the newly established War Gaming Department. (One of the original members of the department was Francis J. McHugh, a civilian, whose War College career had begun in the 1930s and, except for a stint in the army during World War II, would continue until the mid-1970s. Although not as well known to later generations as McCarty Little, McHugh's influence on War College gaming and the War Gaming Department was probably second only to that legendary figure.) Created to operate and maintain the NEWS, the War Gaming Department was charged with developing and conducting not only War College curriculum games, but also with providing support to the fleets as part of the Navy War Games Program.

Established under the chief of naval operations in 1958, the Navy War Games Program was administered by the assistant for war gaming matters (Op-06C). This program focused on analytical gaming, and in particular on the development of computerized combat models. The key technical support groups for the program were originally the Planning Analysis Group of the Applied Physics Laboratory, Johns Hopkins University, and the Warfare Analysis Division and Operations Research Division of the Computation and Analysis Laboratory, U.S. Naval Weapons Laboratory at Dahlgren, Virginia.[61] The Naval War College, and specifically the War Gaming Department and the NEWS, were also included as major elements of the program.

The War College had two principal roles in the program: to provide facilities for the conduct of wargames by operational commanders, and to conduct a course in wargaming for fleet officers. This course was "directed primarily toward operational commanders and members of their staffs . . . engaged in operational planning." Its purpose was "to acquaint them with the Program and to familiarize them with the means by which war gaming can assist commanders in evaluating or rehearsing their plans."[62] This program was the first time "that the Naval War College gaming facilities and personnel were made available to Fleet Commanders for gaming their own specific problems with their own personnel."[63]

The wargaming course, established in 1960, dealt with the **83** history of wargaming and the various types and techniques available. Naturally, special emphasis was placed on the employment of the NEWS to explore fleet operations. Elements of the 1967 course covered "the fundamentals, probability concepts, digital computer simulations, and the capabilities and limitations of simulation."[64] In addition, attendees participated in games to familiarize themselves with the various roles of players and umpires.

The Navy Destroyer School at Newport soon asked the fledgling War Gaming Department to design a game for the use of its students, and the first game was conducted in 1962. These games were designed to let the students practice antisubmarine warfare operations in a very realistic setting. They employed "the proper voice calls and communications procedures" to run their simulated task group, and so gained "an insight into the complexities and uncertainties of the command and control problems of modern antisubmarine warfare" in preparation for the assignments they would soon have in the fleet.[65]

In addition to such fundamental educational uses, the fleet began to use gaming for other purposes. They used wargames to "familiarize the staff with an operations order prior to an at-sea exercise and, as one Carrier Division Commander observed: 'To make our mistakes in the NEWS before we get to sea.'"[66] The fleet also found gaming useful for trying out new tactics and combat formations, testing command-and-control arrangements, and exploring the possible effects of enemy counteractions to plans. The fleet discovered both the value and limitation of gaming because the "Fleet games—like the curriculum games—do not provide formulas for victory, or quick and easy solutions, or even reams of data. But they do provide their planners and players with command-and-control experience and with greater insight into the complex military problems of today and tomorrow. . . . [They also] indicated that the planning and play of NEWS games by experienced naval officers provide valuable contributions to the naval warfare analytical studies sponsored by the Chief of Naval Operations."[67]

In 1962, the traditional mode of playing fleet games locally at Newport was supplemented by a program of remote play. In that year, the commander, Fleet Air Quonset and his staff initi-

84 ated this new capability using secure communications lines to send their operations orders from their own headquarters to be carried out by the umpires at Newport. "The following year a game was played for the Commander, Barrier Forces Atlantic, and then a much larger game known as the Canadian-United States Training Exercise was conducted. This involved six east coast operations control centers and the war gaming facilities of both the Naval War College and the Canadian Joint Maritime Warfare School. A game involving the movement of an amphibious task force to its objective area was conducted in 1964. In this game the commander and his staff played from their flagship which was tied up at its pier in Norfolk. And one year later the first remote-play game was conducted for U.S. and Canadian Pacific Coast Commands. The players functioned from their operations control centers on the West Coast and Hawaii."[68]

Over the next decade, the navy's wargaming program would continue to expand. McHugh cites eight major fleet games played either at Newport or by remote links in 1964 and 1965.[69] But even during this period of growth and expansion, the seeds of change were sprouting both for Newport wargaming and the college as a whole.

FROM NEWS TO NWGS

The big splash made by NEWS, especially its introduction of wargaming to the fleet, helped to elevate wargaming to a level of high interest throughout the navy. Part of this interest resulted from the association of wargaming with the newer quantitative analytical techniques of operations research and systems analysis, which had accompanied the arrival of Robert S. McNamara and his "whiz kids" at the Department of Defense. It was this association that led to the inclusion of both wargaming and analysis in the Navy Wargames Program and would plague wargaming into the future.

It was not long, however, before the flash and promise of the potential of the NEWS gave way to disappointment with its limitations. Some of these limitations were obvious—a restricted area of play, severe restrictions on the number of forces involved, and limited ability to control the speed of the game clock. Others were less obvious, such as the virtual impossibility of adapting the system to newer types of sensors and weapons. Even as the NEWS reached the pinnacle of its reputation and degree of employment, **85** the thoughts of many turned to what steps might be taken to re- place it.

Three options suggested themselves. The analog NEWS system could be expanded to allow a larger playing area and greater numbers of forces to be involved. This option was seen as very costly, and carried no guarantee that it would correct the in- herent deficiencies of the analog system. In addition, both the navy and the computer industry were shifting away from analog systems in favor of digital ones. The second option was to institu- tionalize the ad hoc arrangements developed at the War Gaming Department for using the display capability of the NEWS as a supplement to essentially manual wargaming. Such an approach was certainly inexpensive and familiar to the staff, "but could not offer realism, timeliness of information, or the depth of detail re- quired. Therefore, the complexities of modern naval warfare with its associated exacting logistic implications could not be effectively gamed."[70] This left the third alternative, a shift to the new technological promise of the general-purpose digital computer.

The Warfare Analysis and Research System

The NEWS promise of virtually automatic wargaming now beck- oned from the world of bits and bytes. The new digital computers seemed to hold out the hope of solving all the old problems. "Programs could be specifically structured to game requirements. Real-time information and exact logistic accounting became im- mediately available, and since the ability to change was designed into the system, it should be less susceptible to obsolescence."[71] The vision of the future was given the whimsical acronym WARS, the Warfare Analysis and Research System, and in the early 1970s the War College began to phase out elements of the NEWS and to replace them with pieces of the more modern system, creating a Frankenstein-like hybrid. About this time also, the War Gaming Department acquired the more grandiose title of the Center for War Gaming.

The design philosophy that underlay the WARS was based on the War Gaming Center's perception that the type of gaming central to the Naval War College was best described by the term "decision gaming," and differed from the type of tactical training

86 simulations employed elsewhere. "Decision gaming is a unique educational or analytical process with the primary emphasis upon human choices."[72] To deal with this type of game, the computer program would have to be supremely comprehensive and flexible. It would be founded on a detailed data base of the characteristics and capabilities of sensors, weapons, communications systems, and platforms, as well as on the parameters of platform motion, logistics, and the environment. The Master Simulation Program was to be capable of representing crises or conflicts of any scope or scale that could be located anywhere on the globe. Play could range from one-on-one encounters to a full-scale global war.

The proponents of WARS argued that it had three critically important advantages, unique to the War College system. First was its integrated approach to warfare, melding its models to allow any type of action to be represented. Second, the fact that it was a game system, and not a purely analytical simulation, allowed the easy incorporation of the unexpected decisions and actions of human players into its models. "Beyond the obvious value this offers in the area of decisionmaking experience, this capability permits each model to operate as an entire family of models, each modifiable as the game progresses in order to mesh with specific requirements or permit comparisons of the effect of differing actions."[73] Third, WARS was to be capable of a flexibility in modeling and representation of combat actions unknown in previous systems. The user could select the level of command he would represent, and built-in tactical doctrine would dictate the action of subordinate units outside the scope of his decisionmaking prerogatives.

The WARS ideal was simply described by its War College project officer in the mid-1970s. "The computer assumes the tedious duties of bookkeeping. Most important, however, players and umpires are relieved of the burden of manual labor involving computations, track update, damage assessment, and detection calculations. Unhampered by the minutiae of bookkeeping, the gamer is free to concentrate on his thoughts, his tactics, and his decisions which, after all, is what war gaming at the Naval War College is all about."[74]

As the NEWS attempted to evolve toward this WARS ideal, the War College's hybridized wargaming facilities continued to expand their base of customers and breadth of application. In addition to its ever-present support for the college's curriculum games, the early and mid-seventies saw the Center for War Gaming play host to a wide range of games for an eclectic collection of customers.[75]

- The office of the chief of naval operations, for a series of games to explore advanced tactical concepts, under the direction of the Center for Naval Analyses
- NATO and navy fleet commands, for several regular series of games used to allow staffs to test operational plans and concepts of force employment
- The Naval Materiel Command, which sponsored games to allow the engineers and scientists of navy laboratories to explore the implications of new developments in sensors and weapons
- The Naval War Colleges of the Americas, with an annual game played to explore the routing and protection of convoys in the Western Hemisphere, and which was played with remote links to the South American countries involved; U.S. players were composed of naval reservists whose mobilization billets were with the organizations involved in Naval Control of Shipping.

War College Reorganization

In the midst of this period of evolution and growth at the Center for War Gaming, the War College itself was undergoing another of its aperiodic upheavals. As the navy attempted to adjust to the new philosophy and ways of doing business introduced by Secretary McNamara, the War College slowly began to reorient itself toward the future. The mid-1960s saw the Department of Defense take renewed interest in the affairs of all the service colleges, and civilian officials made repeated calls for a rebirth of professionalism in the military officer corps. One of the tools suggested for use in fostering such a rebirth was the relatively new technique of political-military gaming. (See the next chapter for a discussion of political-military, or pol-mil, games.)

When Vice Admiral John T. Hayward was appointed to the War College presidency in 1966, he began the first of two major periods of restructuring that the college would experience before the middle of the 1970s. He instituted a core curriculum of

88 courses required of all students. He also expanded the number and diversity of civilian faculty at the college by establishing several new chairs, including a Theodore Roosevelt chair of economics and a Stephen B. Luce chair of naval science. The core curriculum and elective courses drew on the expertise of the new faculty to introduce the students to a broad spectrum of military and political affairs. "The final two months of the core curriculum were devoted to enhancing the knowledge of the students in the art and science of naval warfare generally. Specifically, they examined naval capabilities of the United States to engage in worldwide operations in support of national policies. They conducted war games to confirm or disprove their assessments. While the senior level course studied such issues, the command and staff course concentrated on operational matters and especially on the complexities of planning for military operations. Gaming played a large role."[76]

Admiral Richard G. Colbert followed Hayward as president of the Naval War College in 1968. Colbert added a McCarty Little chair of wargaming to those created by his predecessor "to increase the rigor and scope of the expanding games being played on the NEWS."[77] He also expanded the connections of the college with other navies across the world. This work was carried forward and extended by Colbert's successor, Vice Admiral Benedict J. Semmes. Yet by the time Semmes retired in 1972, the college, despite the significant gains of the previous six years, "had neither a stable faculty nor a large, stringently selected student body."[78] Nor had its reputation within the navy improved substantially. In that year, Chief of Naval Operations Admiral Elmo Zumwalt took steps to increase the appreciation of naval officers for the broad scope of the navy's mission. He established a Navy Net Assessment Group to help measure the U.S. Navy's ability relative to its likely adversaries and objectives. He instituted Project 2000 to look at the navy's requirements for the next century. And he chose as the president of the Naval War College a man capable of revising the school's curriculum in ways that would help broaden the navy's thinking. That man was Rear Admiral Stansfield J. Turner.

Turner arrived in Newport with a free hand to make whatever changes were needed to accomplish Zumwalt's principal objectives: "broaden officers who are often preoccupied with their

area of specialty; and help them learn to analyze problems."[79]
Drawing on his own experiences as a Rhodes scholar at Oxford in
1947–1949, Turner sought to develop in the students at the War
College an ability "to reason through problems and to see that
there was more than one answer to any problem. He wanted to
teach them to deal with uncertainty."[80] By ensuring that the students, courses, and faculty of the college met the very highest
standards of excellence, Turner believed that its graduates would
become highly regarded and sought after in the navy, thereby establishing Newport's reputation by results rather than by bureaucratic fiat.

Turner's program focused on the fundamental principles of
strategy, management, and tactics. Students were given heavy
doses of reading in both history and theory. Wargaming continued to play an important role in the educational process, especially in the field of tactics, but Turner's philosophy took it in new
directions. He stressed the use of gaming to help educate individual students, and demanded changes in gaming techniques to
allow more students the chance to command large formations.
The transition to the new WARS system was delayed for two
years to accommodate Turner's belief that students should be able
to "see" the process by which their tactical decisions were evaluated in a game. He wanted a system that would let them "actually
see on a screen how their sonar beam went out. Then they would
see the submarine closing and, as the beam would cross the submarine, the dice roll. On the screen you would see that you had a
50% probability and you did or didn't make it that time. That
would let them understand what probability means. It would let
them see what would have happened if they had used their sonars
in a different way. You could run the action over and over again
and let the man go back and make a different decision."[81] Turner's
focus on education deemphasized the use of the War Gaming
Center to support the fleet. Instead, Professor Jacques Naar, the
first holder of the McCarty Little chair, assisted with the development of more tabletop games to allow more students to take command roles. In Turner's wargames not only would more students
be able to give orders, but they would have a better understanding
of why their orders led to specific game results.

In 1974, Vice Admiral Julian J. LeBourgeois had the unen-
viable task of succeeding the dynamic and sometimes abrasive

90   Turner as president. LeBourgeois's tenure was a time of consolidation rather than revolution. He did, however, make some important changes. One of the more significant ones was the establishment of the Center for Advanced Research in 1975, under Captain Hugh G. Nott, USN (Ret.). Nott had been Turner's chief of staff, and his importance to Newport wargaming is attested to by the fact that the auditorium of the wargaming center at Sims Hall has been named for him. Under Nott's direction the Center for Advanced Research drew on the most promising students as well as faculty and visiting scholars to study important issues, using traditional research techniques liberally spiced with wargaming. In particular, the center led the way in the development of potential employment concepts and tactics for the new Harpoon surface-to-surface missile, explored possible naval operations in the Norwegian Sea, and conducted a study of synthetic fuels.[82]

LeBourgeois also led the way for the next generation of the War College's computerized wargaming system, an evolutionary outgrowth of the WARS concept. In his farewell statement in the *Naval War College Review*, LeBourgeois had this to say about Newport wargaming past and future.

The importance of War Gaming to our officers' professional development, to their decisionmaking ability and to their sense of strategic and tactical awareness is well known. Historically, the College has led the way in the development and the use of gaming techniques. With the new and substantially improved Naval War Gaming facility scheduled, under the direction of Captain Herbert Cherrier, to come on line in 1980, a wide range of new and important capabilities suggests a renewed emphasis on war gaming at the College and throughout the Navy as an aid to professional development. The opportunity for 20 students to be exercised simultaneously as Fleet Commanders will exist. Ultimately, we envisage remote terminals in Norfolk, San Diego, and Pearl Harbor so that Fleet Commanders may test operations plans and develop their subordinates' tactical proficiency without the necessity to leave their headquarters. Terminals aboard ship are also being considered. The future of war gaming at the College will be limited mainly by our ability to use imaginatively the capabilities which will soon exist.[83]

LeBourgeois's vision was on the verge of becoming a reality, but the reality was somewhat different from what he imagined.

The Naval Warfare Gaming System

The Naval Warfare Gaming System, or NWGS, was installed in the War College's new computer system in 1981, and from the first was plagued with minor and major problems. In keeping with the trend that had led to WARS, NWGS was to be virtually a fully automatic gaming system. The computer would keep track of the forces, monitor detections by their sensors, and evaluate the results of attacks. It was designed to support up to 2,000 separate tracks, each of which could represent one or more ships, submarines, aircraft, or cruise missiles. The fact that multiple sensors of several different types (such as radar, sonar, or visual) could be present in each track obviously made for a large number of potential sensor-target interactions. The size and scope of the system, one of its key improvements over the limitations of NEWS and even WARS, proved to be a source of some of its major problems as well.

As originally conceived in the early and mid-1970s, NWGS was primarily envisioned as an educational gaming system, in keeping with the War College's emphasis at that time. Its design, therefore, stressed relatively simplistic modeling of sensor and weapon performance, but also included a provision for adding on more detail in a modular fashion. The plans also included a capability to support the production of postgame reports by allowing easy access to the data base and scenario for a game, as well as the narrative of game events.

As finally installed, NWGS tried to be virtually all things to all potential customers. An advertising brochure published by the prime contractor stressed the system's flexibility, ease of operation, cost effectiveness, high availability, and adaptability.[84] It was supposed to support both educational and research games, ranging from one-on-one tactical encounters to global warfare, and to allow the employment of both existing systems and future projections. In a sense, it was McCarty Little's three-tiered gaming system all rolled up into one main-frame computer and several dozen terminals. It used the same basic structure of several separate command centers for the players and a large "game floor" for the controllers and umpires. Each center and work station now had both an alphanumeric computer terminal and a graphics terminal through which players could access information and give commands.

The mathematical models incorporated into NWGS were designed to simulate the broad spectrum of naval conflict. Separate models, formulated to be as similar as possible, evaluated antiair warfare, antisubmarine warfare, surface warfare, air war-

92 fare, submarine warfare, mine warfare, and amphibious operations. The human players could play against each other, or the computer could be programmed to provide the opposition.

Virtually every aspect of the design, preparation, play, and analysis of an individual game was to be supported by the computer system. A master data base would contain information on all types of weapons, sensors, and platforms, and specific bits of information could be extracted from the data base to be incorporated into a tailored game-design file. Game "preparators" worked with a specialized computer-programming language to choose the level of simulation model used for various game functions, to establish the command and communications linkages, and to define tactical doctrine for the units in play. The actual play of the game could take place at one or more different levels of command. Units not directly controlled by human players operated according to predefined doctrine, a sequence of individual "actions" (such as "launch aircraft" for a carrier) designed to respond to a potential situation likely to arise during play.

Finally, the system supported postgame analysis through its ability to record and display the sequence of game events. It allowed an entire game to be replayed in either real time or at a speeded rate. It also provided for stopping the replay at any point and picking up the play from there.

As with any game or game system, the heart of NWGS lay in its extensive set of mathematical models. The philosophy underlying those models, and in fact the whole notion of automated gaming, were described by the prime contractor in the following terms. "An important advantage of automated war gaming is the capability to simulate activities that occur in real-world situations. With stimuli similar to those expected in actual confrontations, the player continually monitors activity, makes decisions on the basis of the stimuli, and enters decisions into the computer. The computer model controls the simulated activity through complex mathematical routines, making modifications as determined by user inputs.

"Computer simulation models can provide varying amounts of realism and detail depending on the specific requirement. Since war games have different objectives, NWGS provides a family of models at different levels of detail in each of the areas that it simulates." The different levels of detail were the key to making NWGS work. The sponsor, designer, and preparator of each game had to choose from two or three levels of detail in several areas. **93**

- Three levels of warfare area models, simulating the "preengagement activities of forces in the seven basic warfare areas."
- Three levels of kinematics models, controlling the "simulated motion of ships, planes, and submarines according to designed course, speed, and altitude/depth information."
- Two levels of intelligence and communications models, simulating "satellite, HF/DF [high-frequency direction finding] and SOSUS [sound surveillance system] detections, and communications networks that transmit detection information used in games."
- Two levels of detection models, which simulated the "activity of sensors and countermeasures generating appropriate detection and lost-contact alerts."
- Two levels of engagement models, which dealt with the "activities of forces during the engagement phase of warfare, including the firing of weapons and calculating the number, type, and position of individual hits."
- Two levels of logistics models, dealing with "the availability, consumption, and replenishment of fuel, ammunition, sonobuoys, and other designated consumables."
- Two levels of battle damage assessment models, evaluating "the outcomes of engagements on the basis of the number, type, and position of hits."

The choice of which level of model to use was to be guided by the following principal. "A detailed model uses more parameters and more complex computational routines than a more generalized model to provide greater fidelity of simulation. For example, the most detailed kinematics model takes into account acceleration, deceleration, turning, and climb and dive rates in modeling aircraft motion, while the less-detailed model simply performs the requested kinematics change. The degree of detail required for a specific game or for a particular event or entity within that game is selectable. When an event is deemed less important to game objectives, a less detailed model is selected, allowing greater computer efficiency." The problem, of course, was to figure out when and if the more detailed models would be needed. Another problem was to be sure that the models were reasonably accurate, no matter what their level.

The stated goals and lofty promises of the NWGS brochure

94 were not, unfortunately, matched by the system's performance. The actual operation of NWGS was at best a disappointment, and at worst a disaster. In simple terms, NWGS never worked in the manner the Naval War College or the Center for War Gaming expected. It may have been an excellent computer program, but it was an abysmal game and a poor simulation of naval warfare. Some of its more significant problems included:

- A detection model that calculated detection probabilities once every minute and then drew a random number against those probabilities to determine if contact was gained, held, or lost. Such an approach grossly exaggerated the performance of a sonar against a slow-moving target. If, for example, the detection probability were .10, then a detection would occur an average of once every ten minutes—about the same frequency with which a ship might provide updates when it held continuous contact on its target. Sonars simply do not work that way.
- All sensors were assumed to be omnidirectional, and generally provided information that was unrealistically accurate.
- Communications were also too good, with no provision for communications jamming by the enemy. Other types of jamming were handled in a very simplistic fashion.
- No provision was included for the effects of shipping noise on the performance of acoustic sensors; an entire battle group could sail through the middle of a sonobuoy field without degrading the field's detection performance.
- The bulk of the models were designed with the operations of U.S. Navy forces in mind; they had a difficult job of reproducing the ways the Soviet Navy might conduct certain types of operations.

The Center for War Gaming quickly perceived those and other types of problems with the NWGS and began working to correct them. As usual, however, the primary response to the shortcomings of the tool was an increased reliance on the people. Those elements of the NWGS that worked reasonably well were used; those that did not were ignored. Another large and expensive computer system found itself relegated primarily to the role of glorified maneuvering board. Battle-damage assessment, traditionally the term applied to the entire process of determining the results of an engagement and not simply to the effects of individual hits, was carried out by a small team of officers using

paper and pencil, hand-held calculators, and, later, personal **95**
computers.

Once again, a new tool had proved to be less than its promise, but War College gaming continued despite the disappointment. With their problems and shortcomings patched over as
best as possible by the personnel of the Center for War Gaming,
the WARS and the NWGS were still able to play important roles
in two significant developments in navy wargaming in the mid-
and late-1970s, and helped provide concrete examples of what
Admiral LeBourgeois had meant when he called for imaginative
uses of gaming.

The Tactical Command Readiness Program and the Global War Game

Strictly speaking, the Tactical Command Readiness Program
(TCRP) was and is neither a wargame per se nor a Naval War
College project. Newport war-gaming has been a major element
of the program from its inception, however, and so it seems appropriate to discuss the subject here.[85]

In 1975 Admiral Isaac Kidd was appointed Supreme Allied
Commander, Atlantic/Commander in Chief U.S. Atlantic Com-
mand/Commander in Chief, U.S. Atlantic Fleet. The admiral
was concerned about the fact that his principal subordinates, including senior flag officers, lacked extensive real-world experience
in the type of complex military decision-making environment
they would face in a crisis or wartime situation. To help rectify
this perceived shortcoming, Kidd turned to Mr. Erv Kapos, then
a senior analyst working for a defense consulting company, to
help come up with a means of exposing Kidd's subordinates to
stressful decision-making situations in a complex joint operating
environment. Kidd, who had himself participated in wargames at
the Naval War College when he was commander of Destroyer
Flotilla 12, specified only that the program should include some
Newport gaming. (Indeed, many credit Kidd's use and support of
Newport wargaming in the 1970s with saving the program while
the "War Gaming Department wallowed in the WARS/NEWS
morass.")[86]

As ultimately implemented, Kapos's ideas, developed with
the participation of Rear Admiral Paul Peck and others, and also
with the direct involvement of Kidd, led to an approach that em-

96 ployed what Kapos described as "two-dimensional progressive-ness." The quarterly cycles of what was to become the Tactical Command Readiness Program progressed from an orientation to an engagement-level problem, then to a full-scale battle, and finally to a broad campaign. Each step was based on some form of simulation media, progressing from a seminar discussion to programmed instruction, to a one-sided tactical battle problem, and ultimately to a full-scale interactive wargame held at Newport using the War College's computerized gaming facility.

The seminar helped bring the individual commanders and their key staff officers up to speed on the current operational objectives and concepts important to the command, and also helped develop the concepts of operations for the scenario to be investigated in subsequent stages of the cycle. The tactical situations used in the programmed-instruction phase of the cycle allowed the commanders and their principal staff officers to master the technical and tactical information they needed to deal successfully with the more complex elements of their operational responsibilities. The battle problems, held at a navy facility for training the watchstanders of Combat Information Centers, placed the commanders in a realistically simulated combat environment. Finally, the War College game thrust them into the broader and deeper arena of the entire Atlantic theater, and forced them to play against reacting opponents. Through this progressive approach, each of the commanders was given an opportunity to prepare himself fully for the successive levels of activity. By avoiding the dangers inherent in placing a player into an artificial situation for which he was not properly oriented, this approach helped make the most of each phase of the cycle without frustrating, and so losing the attention of, its participants.

Originally a navy-only program, the TCRP soon expanded to include all elements of the Atlantic Command and even representatives from other organizations, such as the Strategic Air Command, the Military Airlift Command, and the Coast Guard. Over the years, despite the deletion or modification of some of its other elements, the Newport game continued to be an integral part of the TCRP. Though at times it wandered away from the interactive, continuous-time game originally espoused by Admiral Kidd into a freer form using irregular time-steps, there has been a recurring tendency for the fleet to try to return the TCRP game to its roots in order to incorporate the stress of decision making under rigid time constraints. Overall, the Tactical Command Readiness Program has been a visionary and usually successful operation, one that has built its success on an appreciation for the critical need of preparing the players, especially flag officers, for their game roles.

In addition to the TCRP games, the mid- to late-1970s saw the development of a second major new use for the wargaming facilities at Newport. Under the inspiration and direction of Captain Hugh Nott, and with the assistance of Commander Jay Hurlburt and Professor Francis J. "Bing" West, the Global War Game (GWG) series sprang into existence in 1979. Designed to explore the nature of a modern global war, Global (as it has come to be known) sought to identify those issues most important in the development of a global strategy, not only for the navy, but for the entire nation. Like no other wargame played by professionals (at least none that we in the West know about), from the very start Global incorporated an incredible wealth of military and political elements, ranging from strategy, tactics, and logistics to diplomacy, international economics, and the role of advanced technologies in future warfare.

The Global War Game has been played for three weeks every year from its inception. During those years it has drawn hundreds of players from all the military services, government, laboratories, and industry. The Global games have employed virtually every gaming technique used at Newport since the end of World War II. WARS and then NWGS served as the basis for the principal naval play. Tabletop techniques, supplemented in later years by microcomputer models, were used to play the air-land battle ashore. Seminar-gaming techniques were adapted to explore special topics in off-line sessions that dealt with subjects like the potential uses of advanced weapon systems, the complications of chemical warfare, or the prospects of nuclear escalation in a conventional war. Players assumed roles as varied as the commanders of task forces, the Supreme Allied Commander, Europe, or the chairman of the Communist Party of the Soviet Union.

THE ORACLE RETURNS?

The first Global War Game was played in 1979. Less than two years later, the trends which had spawned that game, the same

98 trends which pushed senior navy officers and civilians toward taking a broader view of the navy's missions, resulted in another significant restructuring of the Naval War College. In April 1981, at the War College's Current Strategy Forum, Chief of Naval Operations Admiral Thomas B. Hayward announced the forthcoming establishment of a Center for Naval Warfare Studies at the college. In July, former Under Secretary of the Navy Robert J. Murray was appointed to head the new center for the navy's strategic thinking.[87]

The Center for Naval Warfare Studies was built around the Strategic Studies Group, a select team of eight officers, six navy commanders and captains, and two marine corps colonels, specially chosen by the chief of naval operations and the commandant of the marine corps. These officers were picked to represent all the warfighting communities, and they each had extensive operational experience. Their one-year appointment came with a mission—to help the navy develop its thinking about the nature and extent of the maritime contribution to the national strategy.[88]

In addition to the Strategic Studies Group, the Center for Naval Warfare Studies was also to include some of the older elements of the War College organization—the Center for Advanced Research, the Naval War College Press, and the Center for War Gaming. The latter was included at the suggestion of the president of the War College, Rear Admiral Edward F. Welch. Welch "believed the war-gaming center wasn't being used constructively enough for the Navy, and here was a chance to connect the officers who were going to produce new ideas with the officers who had a means of testing ideas and arguing issues on a grand scale through gaming."[89]

Despite some early concern among some of the other elements of the War College that the new kid on the block would keep all the toys to himself, Murray and the Center for Naval Warfare Studies worked to keep up the level of wargaming support provided to the War College curriculum. At the same time, however, they expanded the role of the wargaming center into other aspects of the navy's planning and development processes.

With the strong support of Admiral Hayward and his successor as chief of naval operations, Admiral James D. Watkins, the Center for War Gaming became the official "center for Navy wargaming at the battle group level and above." Its fundamental purpose became the examination of "the larger tactical and strategic issues of deterrence and war fighting." Its new charter was to work with the office of the chief of naval operations, the headquarters of the marine corps, "with the fleets, with the Tactical Training Groups, and with others in and outside the Navy and Marine Corps—other services, OSD [the office of the secretary of defense], the allies—to conduct these larger level games." [90]

Murray initiated changes to revive the spirit and role of the Center for War Gaming. When he arrived in 1981, he discovered that wargaming at the Naval War College was considered little more than a training device for the education of students and fleet staff officers. There were little or no records of the games that had been played in the nearly forty years since the end of World War II except those kept by the Atlantic Fleet. Unfortunately, the turmoil of the early 1970s had prevented the War College from firmly establishing a dedicated group to oversee research, development, and evaluation of games. "One of the real tragedies that resulted . . . was the destruction of all game records held at the Naval War College in 1975. All documentation and histories of games before 1973 were thrown away; with them went all chances for evaluation of games, for some much needed basic research, and even for an assessment of changes in the operational styles and capabilities of successive generations of naval leadership." [91] Recalling the storied past of Newport gaming between the wars and its effects on the navy's strategic planning for the war in the Pacific, Murray was disappointed. He had expected to find a "very lively war-gaming center." What he actually discovered was "an interesting center, but not a lively one." [92]

Murray set about changing things at Sims Hall. "I found that officers at the war-gaming center provided umpiring services but did not themselves have much opportunity or incentive to think through the issues of war fighting on a sustained basis; that the war-gaming center was rarely used to test real war-fighting concepts and war plans, but was mainly for training students; that the games played seldom considered Marine Corps issues beyond their initial movement, or land and air engagements at all, even though it was impossible to get any true measure of the value of naval forces without doing so; and that the war-gaming center

100 didn't work with all the Navy, only the East Coast Navy. These were real problems. Change was needed if war gaming was going to be a useful tool for the Navy in developing strategic and tactical concepts. Something had to be done. Something was done."[93]

In early 1982, Murray and his staff devised a new concept for what the Center for War Gaming was and how it should operate. It would support all the fleets, including the numbered fleets as well as the higher commands. It would "argue real-life questions of strategy and tactics, test real war plans, and develop new concepts of operations." The center would use its own officers "to design the game, write the scenario, analyze the results, and draw the conclusions, and then argue the conclusions anew in subsequent games." The play of the games would be diligently recorded "so the future Nimitzes could always have available a record for pondering, a source of ideas for helping to deter or fight." Murray oversaw the installation of the NWGS and pushed for the incorporation of remote gaming capabilities to link Newport to the fleets and allow the fleet commanders to play games interactively from their separate headquarters. More importantly, perhaps, Murray's team developed a process so that the analyses done by the Center for Naval Warfare Studies and others "could be incorporated in future games, and ideas that emerged in the games could be subjected to outside, or off-line, analysis."[94]

The prime example of this new emphasis on integrating wargaming with other forms of research was seen in the program followed by the Strategic Studies Group. The preliminary research of the officers set the stage for a wargame, which was carefully designed to address the main issues of interest. Players and other participants were selected from a broad spectrum of military and civilian organizations to play a series of games to explore the ideas postulated by the Strategic Studies Group.

In addition to those games, Murray also started the ball rolling on another series of Navy games, the first of which actually took place after he had already been relieved by the new head of the Center for Naval Warfare Studies, Robert Wood. This series of games, known as the POM War Games (for Program Objectives Memorandum, a navy programming document in the annual budget cycle of the Pentagon), began in 1984 under the sponsorship of Director for Naval Warfare Vice Admiral Lee Baggett. The POM games allowed the fleets and the Washington navy to explore jointly the warfighting implications of the navy program within the context of its agreed-upon strategy for future global conflict. The 1988 POM game also explored issues related to low-intensity conflict for the first time in the history of the series.

The POM games were only one example of Admiral Watkins's renewed emphasis on the importance of the Naval War College, and its wargaming facilities, to the navy. He convened several conferences of the navy commanders in chief at Newport, rather than at the more traditional locations in Washington and Annapolis, and on those occasions took the opportunity to conduct wargames with the very highest levels of the navy's uniformed leadership. Watkins stressed the need for wargaming to be closely tied to operations research and tactical analysis. Writing for the practitioners of such research and analysis, the director of the navy's Program Resource Appraisal Division stated that "war gaming has been practiced at the Naval War College for decades. The naval aspect of our current concept, though, is its ready applicability to tactical and strategic analysis and its acceptance by top Navy leadership (as demonstrated by recent CINC war games) as a valued and proven tool. Thus, naval analysts are beginning to make a new kind of impact by using analytical tools to support war gaming and drawing on war gaming results."[95]

Admiral Watkins made a strong case for expanding the role of wargaming in the War College's and the navy's view of the world. Under his influence, Rear Admiral James E. Service, Welch's successor to the War College presidency, took the War College's role as the center of navy wargaming seriously, increasing "the amount of student war gaming three-fold, using crisis action, theater and world-wide games." Watkins himself "developed a direct link for the Naval War College with war gaming, fleet exercises, and operational planning."[96]

That period of the early 1980s was an especially innovative and exciting time for the U.S. Navy. The navy's view of itself and its role in support of the national interest in peace and war changed dramatically. Admiral Watkins articulated an innovative strategy for the use of the navy in wartime, one that stressed early, forward offensive operations as the best defense for the critical assets, areas, and sea lines of communication of the United States and its allies. Newport wargaming, especially the games done for

102 the Strategic Studies Group and the Global series, played an instrumental role in developing the strategic thinking and operational insight that became fundamental to that strategy.

Although Watkins's successor, Admiral Carlisle Trost, has been less vocal about the role of wargaming, there is little evidence of a decline in its stature at the Naval War College or in the navy in general. Fortunately, however, the dangers of wargaming's becoming once again the false oracle in a time of change, while still present for the individual, seems to have been overcome by a broader awareness of the strengths and limitations of the techniques. As Murray said, "If there's anyone who scares me, it is someone who goes to one war game and thinks he's learned the answers to his problem, whatever the problem is. A war game doesn't have that capacity. On the other hand, if there's anybody who frustrates me, it's the guy who thinks he can get nothing out of war games and doesn't want to take the time to bother."[97]

Robert Murray's tenure at the Center for Naval Warfare Studies was an especially important time for the development of Newport wargaming. His sense of wargaming's potential and importance was complemented by his stress on three fundamental elements in wargaming: "good preparation, good intelligence, and good people. People are the most important factor in a good game. Much turns on the quality of officers assigned to the gaming center and the seriousness with which they approach gaming."[98] The navy's high-level support for wargaming and the Naval War College in the early 1980s made the assignment of such individuals more likely than it had been in the past, and the accomplishments of the period bore out Murray's assessment.

Although there have been some changes in Newport gaming since Robert Murray's departure, in many ways they have been changes in form and not in substance. In 1985, the Center for War Gaming reverted to its older title of War Gaming Department. The Naval Warfare Gaming System is being phased out, to be replaced by the Enhanced Naval Warfare Gaming System (ENWGS). But the same sorts of problems the War College has always had with its computerized facilities seem to be cropping up in different guises even with the newer system. Although the final verdict on ENWGS will not be in for several years, its development has again shown the dangers of attempting to develop a **103** massive, special-purpose system that takes several years to design and install, and which is beyond the conceptual grasp of a single designer. Even as the new computer begins its operations, the War Gaming Department finds itself conducting more and more seminar-style games to avoid the strictures imposed by the rigid structure of the computerized system.

From 1972 to 1988, the wargaming pendulum at the Naval War College swung from Stansfield Turner's emphasis on its role in the education of individual officers to Robert Murray's emphasis on its role in educating the navy as a whole. Both attitudes can find many supporters in the pages of War College history. Throughout that long history, the Naval War College has retained its faith in the usefulness and value of wargaming, first as an educational device, and also as a tool to help explore new ideas, and to create and evaluate new concepts and plans. From the cardboard and paper games of McCarty Little to the electronic marvels of the Enhanced Naval Warfare Gaming System, the college has always been open to new tools to assist it in its process of educating and exploring. Sometimes the tool became too important, but never did it remain that way for long. In the end, the secret of the Naval War College's successes with wargaming was perhaps best summarized by Frank McHugh when he wrote that the heart of the Naval War College gaming system was that it was one "in which the *human decision makers*, and not the machine, play the dominant roles." [99]

3

WARGAMING AFTER THE WAR

WORLD WAR II AND THE BIRTH OF OPERATIONS RESEARCH

The previous chapter discussed the role wargaming played in helping the U.S. Navy prepare itself for World War II. As chapter one described, however, wargaming was little used in the West during the war itself. The professional military, wargaming's principal users and sponsors in the United States and Great Britain, had too little time to spend on such theoretical pursuits because they had to get on with the very real business of fighting a global war.

As a result of the military's preoccupation with the practical business of fighting and dying, civilians came to dominate the theoretical study of war, at least among the Western powers. The pioneers were a diverse group of academics who came to be known as operations researchers. They saw their field as "a scientific method of providing executive departments with a quantitative basis for decisions regarding the operations under their control."[1]

Most of the operations researchers were drawn from the ranks of physical scientists. They eschewed the lack of rigorous,

106 statistically valid experimentation inherent in wargaming's emphasis on human decisions. Instead, they stressed the quantitative and scientific analysis of the physical aspects of military operations. They based much of their early work on detailed statistical analysis of operational data, and on the mathematical modeling techniques popular in the physical sciences. In some cases, however, they discovered that wargaming techniques could be useful adjuncts to their more mathematical analyses.

One of the first operations research organizations formed in the United States was the Antisubmarine Warfare Operations Research Group (ASWORG). ASWORG was born in April 1942 under the direction of Professor Philip M. Morse, a physicist at the Massachusetts Institute of Technology. As its name implied, Morse's group focused on the problem of defeating the German U-boats in the Atlantic. Scientists attached to ASWORG developed useful mathematical techniques for analyzing the different aspects of searching for submarines and screening surface forces from their attacks. One of the operations they studied involved hunting for a submarine in a specific area of the ocean in which the submarine was known to be hiding. This type of search could arise as the result of a sighting of the submarine or an attack made by the submarine (a "flaming datum").

One technique employed by Allied forces in such a situation had been simply to flood the area with antisubmarine units, particularly aircraft. These units would attempt to hold the U-boat under water until its battery power or air supply was depleted. When the submarine was forced to the surface, the circling hunters could close in for the kill. It was a simple idea, but it required an enormous number of forces to cover the continuously expanding area of uncertainty in which the U-boat might be located. Many of these saturation hunts were conducted in 1942 and 1943 with but limited success.[2]

ASWORG became involved in the problem, and set about analyzing the vast amounts of operational data available. They soon discovered that the "hunt-to-exhaustion" technique was often failing because "the hunters were exhausted before the hunted."[3] A U-boat might be able to remain submerged for as long as forty-eight hours, during which time searching aircraft would tend to cover too small an area or break off the hunt too soon. Sometimes the weather would interfere, or other missions would pull the searchers off the hunt. "The problem, therefore, was to work out a way for aircraft to regain contact with the submarine that would have a reasonable chance of success without using an unreasonable amount of flight time."[4]

To help in solving the problem, ASWORG devised a wargame to explore the dynamics of the operation. Using specially designed equipment that simulated the ability of the aircraft and submarine to see each other under different conditions, the wargame indicated that the submarine could often spot the plane through its periscope and determine the aircraft's operating pattern, thus allowing the U-boat to optimize its evasion plan.

As a result of this insight, ASWORG developed a new plan, called a "gambit," in which the aircraft would appear to leave the area in hopes of inducing the U-boat to surface and attempt to evade at high speed. If the submarine fell into the trap, it would once again make itself vulnerable to the aircraft, which would have established a new patrol pattern some distance away. Gambit plans were designed in real time to take account of specific operating conditions, and they proved to be quite successful. "Not only were more contacts achieved, but much less flying was required."[5] Thus, wargaming, used in conjunction with operational experience and quantitative analysis, proved a useful tool even in the case of wartime tactical development.

In another such example, at a higher level of operations, the operations research group at the Naval Ordnance Laboratory undertook a study of the possible application of mine warfare to defeating Japan.[6] Progress was slow until the head of the group's countermeasures section, Dr. Ellis A. Johnson, "reinvented war gaming."[7] The group used wargaming techniques to explore the technical and tactical aspects of the problem and also to evaluate the strategic implications. Alfred Hausrath characterizes this effort as "a classic example of the significance of the interrelations between war gaming and operations research."[8] The process went through four stages:

- Technical analyses explored the characteristics of mines and the techniques of their placement that would produce maximum effectiveness.
- Tactical wargames explored different approaches to mining and sweeping specific Japanese harbors.
- Strategic wargames studied the overall implications of mining the Japanese home islands.
- Actual military operations carried out the plans developed

108   with the help of the games and the operational analyses that supported and flowed from them.

An interesting sidelight to the mining study is the fact that Dr. Johnson, the designer of the wargaming approach, was subsequently commissioned as a naval officer and assigned to implement the very operation his research had suggested. In commenting on this unique opportunity, Johnson stated: "As an officer, when I actually conducted the mining campaign against Japan that I had helped to war-game earlier, I found very often that almost every countermove the Japanese made to a powerful mining attack had also occurred in one or more of the war game situations. In fact, the over-all results of the campaign followed closely the results predicted by war gaming."[9]

Despite these and other early attempts to employ wargaming and operations research together, for the most part the operations researchers made little use of gaming. The obvious lack of rigor in wargaming and the apparently scientific nature of operations research led most civilian researchers, who were scientists after all and not soldiers, to rely on operations research and ignore the possibilities of wargaming, if they were even aware of the existence of the technique. Indeed, Ellis Johnson was a geophysicist by training and had been "in complete ignorance of the long and illustrious history of war gaming." Hausrath states: "It was not until later that Dr. Johnson learned that the technique he visualized was well known in military circles but was used primarily for training and testing of plans rather than as a research tool."[10]

THE RISE OF POLITICAL-MILITARY GAMING

Operations research had been born in the dark days of World War II, and most of the groups assembled by the Allies were strictly ad hoc arrangements, which nevertheless became somewhat bureaucratized by the end of the war. The importance and contributions of operations research to the war effort, however, led to a move to institutionalize some of the more prominent wartime arrangements. For example, Admiral Ernest J. King, the commander in chief of the U.S. Navy, urged Secretary of the Navy James V. Forrestal to continue the navy's Operations Research Group (as ASWORG had become known) into peacetime, though at reduced manning levels. Forrestal agreed, and in November 1945 the navy and MIT signed a contract perpetuating the Operations Research Group in the form of the Operations Evaluation **109** Group (OEG). (OEG continues to exist today, providing operational navy commands the world over with analytical support as part of the Center for Naval Analyses.)

In the new civilian-dominated world of military research, wargaming was gradually relegated to a relatively small role as a potentially useful training device and little more. The complex tale of the growth and interrelations of operations research and wargaming in the early postwar period is too intricate and confused to deal with in any detail here.[11] Over time, however, the traditional disciplines of wargaming lost ground even among those who had supported it. The developments in computers and simulation techniques seemed to hold out the promise of doing away with the foibles and unpredictability involved in using human beings as players. Instead, the new methods sought to program the most rational possible decisions directly into the machines, which would quickly calculate the outcomes of many such decisions.

The process of rationalization and dehumanization reached its apogee with the arrival of Robert McNamara as secretary of defense at the beginning of the Kennedy administration. McNamara brought to Washington a new breed of civilian "whiz kids," who preached the doctrine of operations research, systems analysis, and cost-benefits tradeoffs. This new theology buried wargaming beneath a deluge of mathematical analyses and computer simulations. For the systems analysts, buying the right systems required a detailed understanding of the technical nuances of physics and economics, and exploiting new technology, especially to save money, became more important than understanding the nature of war.

In the new environment dominated by civilian analysts, wargaming was all but discarded as a useful tool for conducting research into the physical realities of combat. But the self-evident utility of the gaming idea was taken up by an entirely new breed of postwar defense specialists. Academic strategists and political scientists applied the principles and techniques of wargaming to what was more obviously a less-than-quantifiable subject—political issues and behavior. Thus was born the political-military, or "pol-mil," game, a device that was popularized by the Rand Corporation and spread in surprisingly divergent directions,

110 to the halls of academia and the corridors of the joint chiefs of staff (JCS).

As described in chapter 2, both the Germans and Japanese had experimented with political games during the 1930s, but the applications of the technique in the United States were quite limited before the 1960s. Some American political scientists, notably Harold Guetzkow of Northwestern University, emphasized the use of gaming in theoretical studies of international relations. Such games used role-playing in an attempt "to validate theories about the structure of international politics, . . . [and tried] to quantify the process of international relations by assigning units of national resources, capabilities, and so on, to teams who in a series of somewhat stylized moves attempt to maximize their goals and minimize their losses. . . . [Such games attempted] to simulate interaction not between people but between forces represented by people."[12]

The second school of political gaming focused on more practical goals. Herbert Goldhamer of the Rand Corporation was the pioneer in this type of game, which was soon adopted by academic institutions like the Massachusetts Institute of Technology and Columbia University. Rand's key contribution was to abandon "the attempt to assign numerical values to political and economic factors or to assess in quantitative terms the relative value of alternative strategies. Rand's conclusion was that to formalize the conditions of the game and its payoff would unrealistically oversimplify the real-life world and confuse the assessments of political strategies and tactics which emerge from the game."[13]

One of the early pioneers of academic pol-mil games was Professor Lincoln Bloomfield. Bloomfield had served in the U.S. Navy during World War II, and then in the Department of State for eleven years. By 1960 he was an associate professor of political science at the Massachusetts Institute of Technology and the director of its Center for International Studies' United Nations Project. Bloomfield was an advocate of "the practical uses of the gaming technique for training, research, and hopefully, policymaking. . . . [He stressed games] in which empirical reality and operational values are emphasized."[14]

In 1958 Bloomfield directed an experimental game at MIT's country estate of Endicott House. The scenario revolved around the crisis resulting from a coup in Poland. Such a scenario clearly required strong Soviet, Eastern European, and American teams. But because Bloomfield "anticipated that the problem would go before the United Nations, it was necessary to have teams representing aggregates of countries in Africa, Asia, and Latin America." For three days the game "engaged the rather fascinated attention of a group of senior scholars and officials,"[15] and helped Bloomfield and his associates achieve their principal objective, that of learning more about the gaming technique.

Bloomfield also tested several other potential values in this type of "reality gaming."

- The "attempt to uncover factors in a given operational situation which might not otherwise receive priority or even awareness in more conventional types of planning and research."
- To "help clarify premises which underlie thinking and planning but which are not often if ever put to the actual test of events."
- The "identification of new areas of potential research which may be subsequently investigated by either further gaming or by more conventional research."[16]

Finally, Bloomfield characterized the "notion that the future can be predicted by the use of gaming" as the "ultimate temerity." As he puts it, "Artificial simulation of events is going to distort reality in inescapable ways."[17]

Bloomfield also identified several fundamental factors that seemed important and in common to designing and playing virtually all types of political games. He stressed preparation, and the importance of the choice of the problems as "determining . . . the scope of the game in terms of geography, teams, and players."[18] He also understood the importance of defining the relationship of game time to real time, both in the sense of the time in which the game was set (past, present, or future) and the "clock speed" of play itself.

As to the actual play of the game, Bloomfield pointed out the importance of deciding ahead of time "whether the strategy to be followed by players is to be realistic or what is called a 'deviant strategy.'" He stressed the importance of requiring players "to put on paper their basic strategic goals, their estimate of the situation and their appreciation of how it will probably unfold."[19]

As to the administration of the game, "the role of umpires is vital," as are the "mechanics of reproducing and distributing

112 documents and generally keeping the flow of information moving." Finally "the post-mortem session following the game is perhaps the most valuable event of the political exercise. . . . [If] given adequate time and conducted skillfully, the post-mortem can supply the decisive insights about the planning and action process in the very way in which the game, like the reality it simulates, cannot."[20]

Bloomfield and the other academic gamers were well aware of the limitations of pol-mil gaming and urged that "future uses of the political game need to be approached with prudence." Yet, they also saw two very important benefits to using the games in foreign-policy analyses. "One is the benefit of interaction between several minds—which can, of course, also be achieved around the conference table. The other is the rather more complex set of benefits which flow from the dynamics of the interaction in the form of role-playing, generating a self-sustaining reaction that develops its own course independent of the limits or boundaries with which one starts. There is a potent value in unpredictability, and in exposure to the antagonistic will of another who is operating on the basis of very different assumptions. Neither of these values can be derived from solitary meditation or cooperative discussion."[21]

In the JCS arena, however, pol-mil games became less contests between competing teams and more directed discussions. In a traditional wargame or the type of pol-mil game espoused by the likes of Bloomfield, players were consciously cast in the role of a particular operational commander or staff. The players of JCS political-military games, on the other hand, were explicitly directed to avoid the playing of specific roles. Instead of a player's, for example, donning the mantle of the chairman of the Communist Party of the Soviet Union, an entire team would represent the USSR, in a sense role-playing an entire nation. Because the JCS games often served as the basis for policy recommendations, role-playing was seen as too restrictive and potentially disruptive to a coherent, well-ordered study of the problem.

The mechanics of the standard JCS pol-mil game were described well by Hausrath, and have remained fairly constant from the very first game in 1961. "Two or more teams, each of five or six officials, plan actions and reactions and submit these data through a Control Group of similar size. The teams meet for several hours daily, or on alternate days, to review the scenario and determine objectives, strategy, and plans. Conferences are held with senior officials who come to the game rooms for an hour or so each day. With the concurrence of these senior officials, team moves and strategies are documented and submitted to Control; here they are analyzed in relation to moves from other teams and world influences. Control records the updated world situation and designates the intervening period of elapsed time. The game clock and calendar then are advanced a few hours to several months, and information is submitted to the teams for another cycle of play the following day." [22]

"The initial purpose of the Pentagon games, which involve senior officials like the Joint Chiefs of Staff and Assistant Secretaries of State and Defense, was to improve interagency communication between the new defense intellectuals ("Whiz Kids") . . . and the older professional military staffs." [23] The topics of the games were chosen by the JCS after soliciting suggestions from various government agencies, and a strict non-attribution policy was adopted and continues to this day.

The standard structure and procedure for conducting political-military games was devised in the early 1950s and in many ways reflects an outmoded view of the world. Those were the days of the "bi-polar world," in which the "West" believed itself to be facing a monolithic "Sino-Soviet Bloc" intent on their imminent, and probably violent, overthrow. The pol-mil games of this period naturally took on the standard two-sided form that is still the most popular type today. Such a game structure subtly implies that the only important players on the world stage are the political leaders of the two principal contending parties, usually the U.S. and the USSR. An increasing tendency over the years to require all communications between player teams to take place indirectly through the umpires may have subtly influenced many players to believe "that the posturing, deploying and employing of military forces is the major means available to the two leaderships to try to influence the decisions of their opponents and others." [24] Perhaps this is the reason that all too often one hears discussions of "signaling" the enemy by embarking on a particular course of military action, not only in games, but in the real world as well.

If these sorts of vague, almost amateurish notions of how international politics works are disturbing when we read about them today, they were possibly even more disturbing to the pro-

114 fessional military at the time. Although JCS has conducted pol-mil games almost continuously since the 1950s, other military organizations have tended to view their "touchy-feely" procedures and "squishy" conclusions with well-founded skepticism. As the most prominent examples of research-oriented gaming of their time, it is little wonder that the questionable reputation of political-military games cast a long shadow over all gaming techniques and reinforced wargaming's declining popularity among the active, operationally experienced military.

It seems clear that the virtual castration of wargaming in the 1960s and 1970s in many ways reflected the increased civilianization of military affairs that occurred during the same period. This ascendancy of the amateur over the professional was a familiar, recurring American theme, and one that was soon to make itself felt in wargaming as well.

CHARLES ROBERTS: WARGAMING BECOMES A HOBBY

The postwar period, which saw the increasing involvement and dominance of civilians in matters military, also witnessed the rise of what for the first time might accurately be classified as a wargaming hobby. The founding father of the hobby of board wargaming was, not surprisingly, an American, Charles Swan Roberts II.

In 1952, Roberts was a young man living in Maryland, and he had just been commissioned in the army national guard. He entertained hopes of obtaining a commission in the regular army through a process known as a competitive tour of duty. While waiting for his opportunity, he designed a game that would allow him to "practice war on a board as well as a training field and learn the nuances of the Principles of War in a context that was less noisy."[25]

The game used a rectangular map of an imaginary island inhabited by two contending nations. Over the representation of terrain ranging from flat plains, to forests, rivers, and mountains, Roberts superimposed a square grid, similar to the military system of identifying map locations. To represent the various combat units fielded by the adversaries, he used small square pieces of cardboard on which he drew standard military symbols to distinguish infantry from armor or airborne divisions. Each unit was given a numerical rating for its ability to move through the

squares of the grid, depending on the type of terrain in the square.
Each unit was also rated for its combat capability. Battles were
resolved by comparing the strengths of adjacent opposing units,
rolling a die, and comparing the die roll to a table of results (the
Combat Results Table, or CRT) scaled by the relative strength (or
"odds") of the attacker and defender. Roberts discovered that
playing his game did, indeed, have a tremendous educational
value. But as the Korean War came to a close and the army abol-
ished the Competitive Tour program, his dreams of becoming a
professional soldier waned.

Working in advertising and marketing for a living, Roberts
decided, "[a]lmost as a lark"[26] to publish his game, called *Tactics*,
in 1954. He created The Avalon Game Company (the name
taken from a nearby historical site), and operated out of his basement. The game was printed by a commercial firm and distributed through the Stackpole Company, a publisher of books
on military history and science.[27] From 1954 to 1958 he sold
about 2,000 copies of the game, and "either netted or lost thirty
dollars."[28]

In 1958 Roberts decided to get into game publishing on a
larger scale. To avoid a conflict with another local firm, he
changed his company's name to Avalon Hill. Roberts did not see
Avalon Hill as a pioneer in creating a wargaming market. Instead, he "was convinced that there was a market for realistic
games of a specialty format, designed to appeal to those who enjoy intellectual challenge and prefer competition wherein skill is
a primary virtue." In addition, Roberts and Avalon Hill targeted
the role-playing aspects of their games from the start, realizing
the potential attractions of the idea long before anyone had
coined the phrase. Their early ads challenged the player to outwit
Rommel or Montgomery, Spruance or Yamamoto.[29] His first
games included an updated version of *Tactics* called, originally
enough, *Tactics II*; the first modern historical board game, *Gettysburg*; and the first of his non-war games, *Dispatcher*, a simulation
of railway operations.

Over the next few years, Roberts and Avalon Hill produced
nearly twenty adult games, about evenly split between wargames
and "civilian" titles. The wargames, however, were always the
biggest sellers. In games such as *Chancellorsville* and *D-Day*, the
original square grid was replaced by the hexagonal pattern in

116 common use today. The source for this innovation, one of the most influential devices ever employed in the hobby, was the Rand Corporation.

In the early 1950s Rand had contacted Roberts and in a circumspect manner inquired about the source of the CRT used in *Tactics* (and virtually all of the early Avalon Hill games). Roberts's CRT bore an uncanny resemblance "to the more complex one that Rand was using to wargame World War III and other horrors."[30] This fact was probably somewhat embarrassing to Rand when they discovered that Roberts had devised his table in about fifteen minutes, basing it on the popular military notion that an attacker required a three-to-one superiority in order to be reasonably assured of success. After this encounter with the think-tank wargamers, Roberts became more interested in Rand. Later, he saw a photograph of one of the Rand gaming facilities and noted that they were using an hexagonal grid. This grid allowed movement between adjacent hexagons (or hexes, as they are more frequently called) to be equidistant, whereas movement along the diagonals in a square grid covered more distance than movement across the sides of the squares. Roberts immediately saw the usefulness of this technique and adapted it to his subsequent games.

Although the hexagonal grid was one of the most important contributions Roberts made to wargaming and the hobby during the early years of Avalon Hill, it was by no means the only one. Early games often suffered from a lack of clarity and completeness in their rules, necessitating a hobby "tradition" of postpublication errata. (Certainly a major contribution!) As part of the solution to the problem of inadequate playtesting, which led to such mistakes, Avalon Hill adopted the use of outside testers, drawn from the ranks of the most avid of the budding wargame hobbyists, to identify problems before games were put into production.[31] Other contributions of the earliest games included the introduction of airpower to a land-battle game and the imposition of strict logistical limits on the forces available (seen in *D-Day*); unit-breakdown counters, which allowed a large unit to decompose into smaller subunits (introduced in *Chancellorsville*); and hidden movement and search (used in *Bismarck*).

Roberts also explored the possibility of doing games for the military. Remembering how effective an educational tool he had found the original *Tactics*, Roberts designed *Game/Train*, "the first tactical board wargame, . . . designed to be a squad and platoon level training aid for combat leaders."[32] He attempted to sell this game to the U.S. Army Infantry School at Fort Benning, Georgia. After a demonstration, the military officers seemed to be interested, but the sale was rejected by the civilians who controlled the contracting—perhaps another example of the dichotomy of opinion between the professionals and the amateurs about the utility of wargaming. Despite the failure of his attempt to penetrate the military market, Roberts continued to pursue other noncommercial enterprises. For example, Avalon Hill became involved with the American Management Association in a project to develop simulations to educate managers in new techniques.[33]

Despite the progress and their expanding market, Avalon Hill faced a financial crisis in the early 1960s. In 1963, Roberts turned the company over to Monarch Services, one of its principal creditors. Eric Dott, the president of Monarch, decided that the company could be salvaged. Dott appointed one of Roberts's original associates, Thomas Shaw, to be in charge, and in 1964 a reborn Avalon Hill rose from the ashes.

Although Charles Roberts ceased to be an active force in wargaming after 1963, his influence on the hobby continues even to this day. His name has graced wargaming's equivalent of the movie industry's Academy Awards (the Charles Roberts Awards, or "Charlies"), and his presence at gaming conventions is still a major draw. To put his work into perspective, perhaps it is best to let him use his own words.

The problems of wargame design in the early years were considerable. *Tactics* introduced a totally new method of play which had no parallel in games designed to that point and potential players had difficulty in grasping the simple mechanics. It was revolutionary to say that you could move up to *all* of your pieces on a turn, that movement up to certain limits was at the player's option and that the resolution of combat was at the throw of a die compared to a table of varying results. As simple as this sounds now, the new player had to push aside his chess-and-checkers mindset and learn to walk again. After he learned to walk, he had to master the intellectual challenge of the game itself usually without the benefit of an experienced opponent. The miracle is that the early player managed to play at all, burdened as he was in many cases with poorly written instructions.

The lack of basic skills on the part of the typical purchaser of an early wargame severely limited the designer, forcing him to simplify the play to get the game played at all. Today there is a great body of experienced players who can find opponents with ease and the designer can

118 take wing without worrying about trivialities. Also, today's designer has readily available historic data not so easy to find in the early years. Game design today, enhanced by the evolution of improved graphics, is by no means an easy task, but at least the creator can assume the purchaser will not have difficulty getting the lid off the box.[34]

THE SIXTIES: WARGAMING GROWS UP

With the rebirth of Avalon Hill began a decade in which more and more wargamers would be "getting the lid off" more and more boxes. Over that next decade or so a small but continually growing body of civilian (and military) hobbyists began to enter the world of gaming. As the number of players grew, clubs formed and a gaming press began, initiated and dominated by Avalon Hill's house organ, *The General* (begun by Thomas Shaw in 1964), but including a varied array of mimeographed newsletters published by the clubs or dedicated individuals.

One of the first of Avalon Hill's new games in the post-Roberts era also began a new trend in the research involved in game design. Following the publication of *Afrika Korps* in the spring of 1964, work began on *Midway*. As part of a promotional scheme for the game, Dott contacted Rear Admiral C. Wade McClusky, an authentic hero of the Battle of Midway (he was the commander of the dive-bomber squadrons of the USS *Enterprise*), who happened to live in a nearby Baltimore suburb. McClusky was only too glad to endorse the game, especially after Avalon Hill asked his permission to print his previously unpublished eye-witness account of the battle, which he had penned immediately after the events of June 1942.

The fact that McClusky's account did not completely agree with the official navy records or the official history "set a radical change in the course of Avalon Hill research and design. Henceforth their commitment was to spend more time developing primary sources rather than rely on the more easily obtainable secondary sources."[35] McClusky became the charter member of Avalon Hill's Technical Advisory Staff, which later came to include General Anthony McAuliffe (U.S. Army commander at Bastogne during the Battle of the Bulge) and Colonel Donald Dickson, a marine corps veteran of Guadalcanal. Despite Avalon Hill's recruitment of such illustrious figures to help with some aspects of the research for their games, they achieved only mixed success in their attempts to produce more historical accuracy.

The new-found drive for greater realism, however, was to be- **119**
come a major theme for the future development of the hobby.

One of the principal navigators into these uncharted waters
was to be James F. Dunnigan. During the early 1960s, Dunnigan
wrote Avalon Hill several letters commenting on the sloppy historical research apparent in some of the early titles. It was not
until 1965, however, that Dunnigan finally met Thomas Shaw.
Doing research for a college term paper dealing with the accounting aspects of the toy business, Dunnigan showed up at the Avalon
Hill display during a toy-industry show at New York's Toy Center
and began asking questions of the first body he saw, which just
happened to be Shaw. The result of that conversation was an "A"
on the term paper and an invitation to visit the Avalon Hill offices
if he were ever in Baltimore.

Over the next year, however, Dunnigan launched his own
amateur publication. With help from fellow wargamer and history buff Vic Madeja, Dunnigan began writing *Kampf*, a series
of short pamphlets devoted to detailed historical analyses of important campaigns of the Second World War. In March 1966
Dunnigan made a trip to Washington, D.C., to gather information for the series, and he stopped in Baltimore to renew his acquaintance with Shaw.

During their conversation, Shaw asked Dunnigan if he knew
anything about the Battle of Jutland. In classic style Dunnigan
knew just enough, as he puts it, to fake it, and he was more than
a little surprised when Shaw asked him if he would like to design
a game on the battle for Avalon Hill. Although Dunnigan had
little familiarity with naval affairs, and the thought of designing a
game for publication had never entered his mind, he did not
hesitate to answer yes.

Dunnigan saw the deal as an outgrowth of his record of historical research and his insight into the marketing aspects of the
business. For Shaw, however, Dunnigan got the job "by questioning Avalon Hill's R&D methodology to the point where he
was given a design contract" just so he would stop being a pest. [36]

Dunnigan and his questions reflected the evolution of the
wargamer as a player of games into the wargamer as a student of
history as well as the games it inspired. From the initial fascination of just having a game that could provide a taste of combat
command, players began more and more to ask why the rules
were the way they were. From the earliest debates about the ac-

120 curacy of the orders of battle for Avalon Hill's original *Battle of the Bulge* to the primitive attempts at revising the rules of the older games to reflect their own interpretations of historical factors and events, the players of games were quickly turning into fledgling game designers. And as designers, they were beginning to ask why not when it came to new ideas and innovations.

Tom Shaw's contract with Jim Dunnigan was the source of Avalon Hill's first free-lance game design, and Dunnigan was the first serious example of gamer-turned-designer. The end product of the collaboration was the game known as *Jutland*, published in 1967. Unaware of the history of naval miniatures-gaming or of Fred Jane's wargame, Dunnigan had basically "reinvented the wheel" and constructed his system from scratch, discovering many classic solutions to the problems of tactical naval wargaming.

"*Jutland* was a radical departure from the norm. It was the first Avalon Hill game to borrow heavily [albeit unknowingly] from miniatures. It was the first to dispense with the traditional gameboard. It was the first to print a four-color process painting."[37] Its design was a breath of fresh air into a slowly developing, if not quite stagnant, art of game design. *Jutland* dealt with the largest fleet engagement of World War I, which took place between the British Grand Fleet and the German High Seas Fleet in May of 1916 off the coast of Jutland (Denmark). The game was played on two levels. First, the players planned the steaming order and deployment of their fleets, and plotted their movements on a paper map of the North Sea. (Reproduced in figure 5.) By superimposing the plots, the players could determine whether any elements of their forces had come into contact. If so, play transferred to the floor or any other large flat surface, and the players laid out their ships in the predetermined steaming formations. The ships were represented by one-and-one-half-inch-long rectangular counters depicting an overhead view of each individual capital ship and simple silhouettes for groups of smaller vessels. Using ruler-like devices provided in the game, players maneuvered their ships into range of the enemy and then fired their guns. Results of gunfire (or torpedo attacks) were determined by referring to detailed tables, and hits were marked off against each ship in terms of the number of main guns put out of action. When the full battle was in progress, *Jutland*

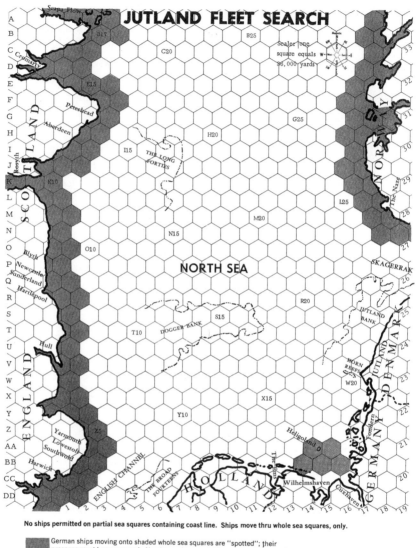

No ships permitted on partial sea squares containing coast line. Ships move thru whole sea squares, only.

German ships moving onto shaded whole sea squares are "spotted"; their presence must be announced. Number and types of ships need **not** be stated.

German Mine Fields — only German ships may enter

Figure 5. The Fleet Search Map from the Avalon Hill Game Company's *Jutland Game*. (Jutland Fleet Search Map reprinted with permission by the Avalon Hill Game Company. All rights reserved.)

122 was an impressive spectacle. (Indeed, when, as an undergraduate at Duquesne University, I spread the Grand Fleet out on the floor of the Student Union Ballroom, I attracted a great deal of attention, and a few new opponents!) Unfortunately, it was not a big seller, probably because of the inordinate amounts of space (and often time) it required to play. *Jutland* was, however, a critical success.

So encouraged, the formula was put to the test yet again. In 1968 Avalon Hill published Dunnigan's second design. The game *1914* dealt with the opening months of the First World War in France and Belgium. The four-color printing process, which had graced the box-cover art of *Jutland*, was now applied to the mapboard of *1914*. Dunnigan used a "step-reduction" system (in which units that suffered losses in battle were replaced by counters with reduced combat strength) to reflect the severe attrition so characteristic of the popular perception of World War I combat. He also included elaborate strategic-level rules to allow the Germans to decide how many forces they would commit to the Russian Front, rather than restrict them to the historical levels. *1914* was unquestionably the most complex game designed to date, but it "was a great sales success."[38]

That success was to be nothing, however, when compared to the sales of the next joint Avalon Hill–Dunnigan venture. In May of 1969 Dunnigan and a friend by the name of Redmond Simonsen approached Shaw with a prototype game called *Tac Force 3*. The game was a low-level tactical re-creation of combat on the Eastern Front in World War II, using units representing platoons, companies, and batteries of tanks, infantry, and guns. Shaw changed the name of the game to *Panzerblitz* and with its publication in 1970 irrevocably changed the shape of the wargaming hobby.

Everything about *Panzerblitz* seemed new and better. Instead of a single large but unvarying mapboard, *Panzerblitz* used three smaller mapboards that could be joined together in any combination, either side to side or end to end. This "geomorphic" style of map, which had been used earlier in the year in the disastrous and truly awful game of *Kriegspiel* [sic], would become almost *de rigeur* in future tactical games. The counters, too, were among the most attractive and "sexy" yet produced. They were about 3/4″ square as compared to the standard 1/2″ size of other games. Instead of dry NATO-style unit symbols, the vehicular counters were adorned with striking silhouettes. Gamers were en- **123** thralled by the mystique of the German *Tiger* tank and the Soviet T-34/85. Instead of abstractions, combat was now carried out by ranged fire and "close assault." No longer would players be restricted to acting as generals far removed from the real fighting; *Panzerblitz* brought the action up close and personal for the first time. Unfortunately, events half a world away were developing in a similar vein, only the action was taking place in Southeast Asian rice paddies and not on an American game board.

VIETNAM: WARGAMING IN DECLINE

Just as World War I had dampened the popularity of Fred Jane's game and World War II had pushed Fletcher Pratt's game into the closet, the war in Vietnam had profound effects on both hobby and professional games from the mid-1960s to early 1970s. The professional political-military games seemed to be little more than amateur and grandiose versions of "King of the Hill," and many of the games played specifically to explore the Vietnam conflict were notoriously flawed and misleading. Furthermore, the games and simulations the professional analysts used to explore questions of nuclear war left a bad taste of callousness and Strangelovian disgust in many mouths. Gaming's reputation among defense professionals, including finally even the civilians who had adopted and misused them, was at low tide.

One of the most important topics for Defense Department wargames in the early 1960s was the continuing bugaboo of nuclear (or in those days "atomic") war. Significant activity on this subject had begun during the Eisenhower administration, when the air force adopted the Rand Corporation's Air Battle Model to conduct a net assessment of U.S. and Soviet nuclear warfighting capabilities.[39] This game was played by Blue and Red teams in the summer of 1955 at the Air War College at Maxwell Air Force Base, Alabama, and was somewhat notorious for the inability of its design to deal with the attack and defense of navy carrier battle groups. The results of the game indicated the potential devastation of a strategic exchange; it also showed "that senior government officials are sometimes not equipped to interpret faults in games and thus can be misled by wargame outcomes."[40]

Nuclear games of the 1960s evolved around the interplay of the U.S. Single Integrated Operations Plan (SIOP) for strategic strikes against the USSR, and the best U.S. guess about how the

124 Soviets might employ their weapons, the so-called red Integrated Strategic Offensive Plan (RISOP). These plans were typically played against each other in computers with little actual game play. On the procurement planning side, another Rand game called SAFE (Strategy and Force Evaluation) was used to explore alternative strategies.[41]

As the Johnson administration got underway, however, more and more JCS games began to revolve around the slowly building situation in South Vietnam. Sigma I-64, for example, explored U.S. options for secret campaigns to respond to North Vietnamese aggression, including "a secret bombing campaign."[42] Deputy Under Secretary of State Seymour Weiss criticized this approach because of the virtual certainty that the campaign would be exposed through the normal workings of the press or the capture of a U.S. airman involved.[43]

Less than four months after Sigma I-64, on 2 August 1964, the Tonkin Gulf incident gave the U.S. its entree into direct combat activity in Vietnam. A month later the Joint War Games Agency of JCS began play of Sigma II-64.

Sigma II-64 used an unusual structure in which senior policy-level teams oversaw and directed the activities of the lower-level working teams. Senior-team discussions focused on broad questions like whether or not tactical nuclear weapons should be used, whether partial mobilization should be declared, and whether the Nationalist Chinese army should be "turned loose" in Vietnam. The objective of the game was to explore alternatives "to compel the enemy to cease support of insurgencies, to assist local forces as necessary in the elimination of the insurgents who thereafter persist, to reunify Vietnam and to achieve the independence and security of friendly nations in the area."[44]

Many of the alternatives considered and adopted by the Blue players of Sigma II-64 were also to make their appearance in the real world: the introduction of ever-increasing numbers of U.S. forces, escalation of bombing in North Vietnam, and planning for amphibious operations.[45] But while the broad outlines of the coming years can be seen in Sigma, the details of the war eluded other gamers focusing on narrower issues.

A great deal of this lower-level gaming was done by the Research Analysis Corporation, a major force in analysis and gaming during the 1960s and 1970s.[46] One of the principal tools in their Vietnam work was an adaptation of an older model, THEATERSPIEL, which had been developed as a theater-level analytical game for studying conventional warfare.

The THEATERSPIEL Cold War Model was designed "to simulate conditions of insurgency in Vietnam and effect an orderly relation between military and non-military factors."[47] A paradox if ever there was one! The model combined various elements of the THEATERSPIEL computer system (intelligence, military, and logistics models) with a new model of "the economic, political, psychological, and sociological aspects of cold war . . . [and] terminal models [to] evaluate the output of the other models and determine political changes."[48]

The model "was built on the objective of winning the support of the population of an area for political authority (Red or Blue) while assuring security and contributing to the economic and social development of that area," in other words, the same flawed assumptions that underlay the U.S. "pacification" program.[49] To make matters worse, the model assigned "arbitrary values to depict the impact and effects of military and non-military personnel in the political, economic, and psychological-sociological fields."[50] Little wonder that such games failed to prevent the long series of frustrations of U.S. military and political strategy in Southeast Asia.

Little wonder, too, that the games (and let us not forget all the other "hard" analysis) of the Vietnam era, and the decisions that those games may have affected, could lead critics like Andrew Wilson to write the following words in 1968.

Could one be so sure, I asked myself, that the American military planning apparatus might not in the end be vindicated, even in Southeast Asia?

Seven weeks later . . . I found myself on operations with the U.S. Marines near the Vietnam Demilitarized Zone. In now different circumstances, watching the parachute flares sink down on the sad hillsides, I asked myself the same question again. I was seeing, not for the first time, the lessons of war games applied in action—and some, I had to admit, had been well and profitably learned. The logistic apparatus in Vietnam was superlative. I had seen operations by the Air Cavalry that were as perfect in execution as a battle school firepower demonstration. . . .

It was only when one looked at the toll of civilian casualties, the impoverishment of the countryside, the growing refugee problem, the

126 degradation and demoralization of Saigon, that one saw the extent of the moral and strategic trap into which America had fallen—a trap from which even "victory" could never rescue it. Here were the factors with which no war game had reckoned, or perhaps could *ever* reckon. The consequences of overlooking them—the cost in life and treasure, the loss of allies, the exposure of military impotence, the effects on American national unity and the American character—would reach out in ever-increasing circles for years to come.[51]

The JCS decision to avoid (or at least not condone) individual role playing, and the push to reduce the role of human players in so-called "analytical" games, undoubtedly contributed to the self-deluding tendencies exhibited by many of the games played about the conflict in Vietnam and other potential trouble spots even today. From its very start, gaming had drawn much of its value from the tension between opposing intellects and the competitive instincts of the players. The competition could be especially keen when a player was asked to step into the shoes of a prospective or actual real-world opponent. In that case, in trying to outwit the other guy the player would actually be attempting to defeat his own "side" in the real world, in order to "win the game." Often such situations produce exceptionally insightful play, as the "opposition" delights in making their real-life colleagues squirm. The JCS approach removed much of this sharpening of the competitive drive, and with it went many of the incentives and abilities of the players to think well beyond the limits of "conventional wisdom" and the dictates of the "rational actor" theory of political behavior. In real games, as in the real world, people sometimes act irrationally (at least from the other side's perspective), such as when they are angry, or confused, or pressed by time. The carefully structured, non-role-playing, almost pristine nature of some JCS-style games removed many of those pressures and perspectives and so produced unrealistically well-ordered decisions.

In an attempt to salvage the most useful elements of their tools from the reaction of disillusioned users and outspoken critics, professional practioners of wargaming and simulation tried to present a cautious and balanced view of the technique's strengths and weaknesses. Hausrath warned: "Enthusiasts, impelled by visions of gaming and simulation potentialities, may try too much, too quickly, too soon."[52] He agreed that the art of wargaming was making great advances, but also saw that its "limitations have been reduced in degree but have not been eliminated."[53] In a nutshell, the fundamental problems with creating the "perfect" wargame, one capable of predicting the future course of battles and wars, lay in the fact that "man's understanding of the process of warfare is incomplete and inadequate. . . . Moreover, any simulation, model, or war game is incomplete when measured against all the factors involved in a real combat situation."[54]

To many inside the defense community, and even more so to outsiders, the majority of professional wargamers were perceived almost as "mad scientists" attempting to lead the world to its doom to prove their pet theories. Books like Wilson's seemed to reflect the public mood when it condemned the evils of wargaming as an academic and antiseptic way to plan to send men and women, or even whole societies, to their deaths.

Not surprisingly, the wargaming hobby also found itself experiencing its own crisis during the Vietnam era. Wargaming's apparent glorification of war as an exciting and fun way to spend an afternoon seemed incongruous to many of the young adults on whom the hobby counted for its membership and support. It was hard to reconcile playing at war when so many of their contemporaries and friends were dying in the reality of Vietnam.

Some gamers restricted themselves to wargaming with miniatures. Although modern-era miniatures games did exist, the bulk of play and interest took place in older, simpler times. Games dealing with the battles of Caesar, Nelson, or Napoleon could take advantage of the color and pageantry that miniatures could produce, thus avoiding the more unpleasant associations of modern tanks, planes, and warships. Furthermore, much of the interest and pleasure of miniature gaming was centered around the miniatures themselves. Acquiring interesting figures, weapons, or ships and painting them authentically was almost as important, and in many cases even more important, than playing games with them.

As the modern miniatures-gaming hobby developed, it did so in a fairly self-contained way, at least in the United States. There were, and are, few major, professional, national publications dealing with the miniatures hobby, *The Courier* being one of the better known. Indeed, one of the most interesting and academic of all gaming publications is an amateur-quality magazine

128 called *Empires, Eagles, and Lions. EEL*, as it is known, focuses
on the warfare of the late-eighteenth and early nineteenth centuries, and particularly on Napoleonics. Boasting contributors
such as David Chandler and Paddy Griffith of Sandhurst, *EEL*
has achieved a high reputation for scholarship, but is virtually
unknown outside a small audience of miniaturists.

Although it is difficult to document, the likelihood is that
the majority of miniatures rules in use today have never been
published. They have been written and used by hundreds of local
miniatures clubs and informal groups. Despite advances in technique and improvements in publicity, especially the boost it
received from its link to the popular science-fiction and fantasy-
role-playing games that developed in the 1970s, American miniatures wargaming never broke out of its Vietnam-era period of
limited, relatively low-key appeal. Such was not the case with
board wargaming.

THE REBIRTH OF WARGAMING

In the midst of wargaming's decline, the seeds of its rebirth had
already been sown. Among professionals, the military adjusted to
the McNamara revolution. Programs of higher-level military
education began to produce quantitatively trained officers capable of carrying out their own systems analyses, and also capable
of competing with, and criticizing, civilian analysts on their own
ground. The services developed an increasing appreciation of the
fact that much of the earlier civilian-driven analysis was perhaps
rigorous but almost certainly wrong because the analysts lacked
any real understanding of actual warfare. As the military professionals began to assert the importance of combining operational
experience and an appreciation for the realities of the field with
the theoretical and mathematical techniques of civilian operations researchers, wargaming began to make its appearance once
again as a valuable tool for demonstrating and exploring the implications of that experience in the research arena. The new electronic techniques pioneered at places like the Naval War College
made the processes of gaming relatively less time consuming, and
less obviously subjective.

At the same time, the dominance of Avalon Hill over the
board-gaming hobby was broken by the increasing influence of
James Dunnigan and Simulations Publications, Incorporated.

SPI, as it came to be called, grew from the seed of *Strategy &
Tactics* magazine, founded in 1966 by Christopher R. Wagner
and Lyle E. Smethers.[55] Wagner, then an air force staff sergeant
serving in Japan, was dissatisfied with the limited focus and uneven quality of Avalon Hill's magazine *The General*. Looking to
include articles dealing with the broad spectrum of wargaming,
including miniatures and the small but growing number of non-
Avalon Hill games, Wagner began to publish *Strategy & Tactics*
on a virtual shoestring. After seventeen issues, the shoestring
broke, and Wagner went shopping for someone to take over the
magazine. He found James Dunnigan, who had been toying with
the idea of publishing a large number of games, but had been
unable to find the proper outlet for them. Seeing *Strategy & Tactics* as just such an outlet, Dunnigan decided to take over the
magazine.

Avalon Hill had always looked on wargames as games, something people might be willing to buy in small quantities for the
sake of amusement or, as was the case with chess, to study and
master. In keeping with this philosophy, they had followed a policy of producing very few new games yearly, and many of their
games, especially the earlier ones, used the same basic system of
play. This policy helped ensure that most of the gamers of the
early 1960s were familiar with most of the games then in existence. Furthermore, new games were relatively easy to learn, because they were all so similar. Such factors were in fact critically
important to the initial growth of the hobby, when players were
few and introducing new people to gaming required overcoming
the obstacles of "long" rules (four to eight pages) and playing
times of more than two hours.

Holding onto an almost unique vision, Dunnigan saw games
as more like books, each with its own particular perspective (and
therefore its own specialized game system), and drawing on a potentially enormous number of different topics. More importantly,
from the practical perspective, they were things that people might
be willing to buy in quantity, if the price were right, to read or
play only a few times, with an opponent or solitaire, to draw what
lessons they might, and then to go on to further exploration of
the same or different topics. When Dunnigan took over the reins
of *Strategy & Tactics*, he found a ready-made forum from which
to test the validity of his ideas. Dunnigan's fertile imagination en-

130 abled him to produce just the sort of new and innovative designs the gaming public thirsted for. One description of Dunnigan's "genius" for game design characterized it as "his refusal to be bound by a fixed pattern when that pattern does not fit."[56]

In a surprisingly short period of time, Dunnigan collected around himself a group of people willing and able to experiment with new gaming concepts, and to explore new, obscure, and diverse gaming topics. These second-generation wargaming pioneers included Redmond A. Simonsen, a graphics artist with an abiding interest in wargames, who as art director of SPI led the way in integrating more and more of the hard data required to play a game directly onto the map and counters. He also carried the banner of standardization of gaming terminology and symbology. Some may argue, with more than a little justification, that Simonsen's utilitarian view of gaming graphics glorified pure functionalism over the later and more appealing approach that balances the practical and the aesthetically appealing. Yet, the combination of Dunnigan's urge for innovation with Simonsen's drive to make the innovative at least look and feel familiar proved a dynamic and influential synergism that helped wargaming develop by quantum leaps.

While professional gamers struggled during the late 1960s and early 1970s to devise computer-based approaches that could adapt old ideas of gaming and analysis to the new systems of warfare, the wargaming hobby exploded with a wealth of new ideas for exploring old (that is, historical) systems and methods of warfare. Over the next decade, SPI produced hundreds of games. Scores of those games were eminently forgettable, but scores more were brilliant in all or some facets of their design. They explored the full spectrum of warfare, from ancient times to the far future, from man-to-man combat to the clash of whole armies and nations. Logistics, command and control, morale, administration, and the vagaries of unpredictable ("random") events were introduced as the first truly professional designers of hobby games began to define their trade.

Dunnigan and SPI also achieved one of Charles Roberts's long-held and unfulfilled dreams. For a price of $25,000, the U.S. Army contracted with Dunnigan and SPI to produce a board game that could be used as a training aid for infantry units. The following words from SPI's *Moves* magazine describe the origins of this game, published for the hobbyist under the title *Fire-fight*: "One rather obvious reason for the Army wanting the game is so that it can 'play around' with their tactical concepts on paper before taking a lot of troops out to do it in the field and, of course, before they have to do it for real. For civilians . . . the main purpose of the game is to let people know what is going on in the military. Hopefully everyone will get something out of it." [57]

In games like *Firefight, Red Star/White Star, Mech War '77, Fulda Gap*, and others, Dunnigan and SPI lead the wargaming hobby into new realms, not only of game design but of gaming topics. No longer would games deal only with history or science fiction. In these "future history" games, SPI was introducing, to hobbyist and professional alike, the utility of gaming to help civilians better understand the potentially violent world in which they lived.

GROWTH AND CHANGE

Dunnigan and his cohorts at SPI, including John Young (*La Grande Armée, Dreadnought*), Irad Hardy (*Firefight, War Between the States*), Frank Davis (*Frederick the Great, Wellington's Victory*), Richard Berg (*Terrible Swift Sword, Conquistador*), David Isby (*Soldiers, Air War*) and others, injected new life and new ideas into the wargaming hobby. The success of SPI proved that the hobby could support both diversity and quantity. The free-lance designers, led by the likes of John Hill (*Bar Lev, Squad Leader*) and John Prados (*Pearl Harbor, Third Reich*), began to play a more influential role, even as SPI and Avalon Hill developed design staffs of their own. In addition, small companies and even individual game producers sprang up. Some, like Robert Bradley and his exciting game *Alesia* dealing with Caesar's decisive battle in the Gallic Wars, or Roger Cormier and his legendary *Trafalgar*, disappeared after publishing one or two "labors of love." Some, like Game Designers Workshop and their legendary *Europa* series of games dealing with all of World War II in Europe, survived and grew. Others, like Battleline, produced limited quantities of high-quality games like *Wooden Ships and Iron Men, Shenandoah*, and *Flattop*, many of which were later purchased by Avalon Hill to refurbish and give a wider distribution. These small companies made up the so-called "third world" of a wargaming hobby dominated by the rivalry between

132 Avalon Hill and SPI. Meanwhile, out on the fringes lurked the world of fantasy-role-playing (or FRP) games, devised by a group of board wargamers led by Gary Gygax and David Arneson.

By the mid- to late-1970s the new strength and diversity of the hobby was presenting potential game players with a new problem. It was no longer a matter of waiting with bated breath to see what one or two new games Avalon Hill would produce this year. Now the question was which of the hundreds of games available should the hobbyist spend his money on? Indeed, in 1977 SPI's book *Wargame Design* listed over 500 games and more than 25 game publishers. Another 1977 book, *The Comprehensive Guide to Board Wargaming* by avid British wargamer Nicholas Palmer, undertook to give potential wargamers a list of "every professionally produced wargame"[58] that the book's author knew to be available for purchase (or scheduled to be published) at the time. Each game was summarized in a few lines, and its ratings from reader polls published by both Avalon Hill and SPI were also included. The list numbered nearly 300 games, and the top games of the time, based on polls conducted by SPI and Avalon Hill, are shown below.

Top Games of 1977[59]

From the Simulations Publications Incorporated Polls (out of 202 games):
1. *Drang Nach Osten* (Game Designers Workshop)
2. *Bataille de la Moskowa* (Martial Enterprises)
3. *War in the East* (Simulations Publications Incorporated)
4. *Frigate* (SPI)
5. *Antietam* (SPI)
6. *Wooden Ships and Iron Men* (Avalon Hill)
7. *Dreadnought* (SPI)
8. *Bar Lev* (Conflict Games)
9. *Jena-Auerstadt* (SPI)
10. *Torgau* (GDW)
11. *Crimea* (GDW)
12. *Sinai* (SPI)
13. *Shiloh* (SPI)
14. *Chickamauga* (SPI)
15. *Wagram* (SPI)
16. *Global War* (SPI)

17. *Kingmaker* (Philmar Limited/AH)   **133**
18. *Chinese Farm* (SPI)
19. *West Wall Quad* (SPI)
20. *Blue and Gray Quad I* (SPI)
21. *Wurzburg* (SPI)
22. *Borodino* (SPI)
23. *Napoleon at War Quad* (SPI)
24. *Battle of Nations* (SPI)
25. *Narvik* (GDW)

From the Avalon Hill Game Company Polls (out of 25 Avalon Hill games):

1. *Wooden Ships and Iron Men*
2. *Anzio*
3. *Panzer Leader*
4. *Richtofen's War*
5. *1776*

The numbers and diversity of the new games and new game designers cried out for a means to help make the consumer aware of which games were worth buying and which could be passed up. In addition, as games became more and more complex, the debate over the balance of realism and playability, which was as old as the hobby itself, took on greater urgency. Players needed to know what they were getting for their hard-earned dollars. For some, relatively simple and exciting games were what they wanted. For others, intricate detail and technical accuracy were the ideals. Gamers wanted more help in identifying the games that best suited their tastes. This need was met by the growth of an active and professionally produced hobby press.

Avalon Hill had led the way in professional hobby magazines with the introduction of *The General*, and *Strategy & Tactics* had initiated the notion of an "independent" gaming voice. When Dunnigan took over *S&T*, that magazine had lost much of its reputation for independence, despite the fact that, unlike *The General*, it occasionally mentioned games produced by other companies and sometimes even spoke very favorably about them. Dunnigan had soon expanded SPI's horizons, however, and in 1972 it began to publish a second magazine, *Moves*, to deal specifically with game-design concepts and theory. The short notices and reviews of games that had appeared in *S&T* shifted largely to the new magazine and expanded in scope and depth. This game-

134   review function, which had been a staple of many of the amateur and club magazines of the 1960s, had so increased in importance that it inspired the creation of a new professional-quality magazine devoted precisely to presenting independent critical reviews of new games from all publishers.

Fire & Movement magazine, subtitled *The Forum of Simulation Warfare* (later, *The Forum of Conflict Simulation*), was founded in 1976 by Rodger MacGowan, its editor, director, and guiding hand. The first issue revealed MacGowan's motivation and philosophy: "Some readers will ask why *Fire & Movement*, when there are so many other wargaming magazines. The answer is quite simple—to date, the hobby still does not have a magazine with the capacity to cover the entire field. There is no single major publication able to cover games published by *all* the game companies. This situation constitutes a *need* and *Fire & Movement* will work to fill this need in the hobby. Our publication is designed to provide coverage in an independent fashion. *Fire & Movement* publishes no games. We have no games to 'push.' Our concern is with the hobby and *you are* the hobby."[60]

Fire & Movement contained review articles by many well-known gaming personalities, and in a major innovation, also included responses to those reviews by the game designers themselves. Although the appearance of *F&M* "was greeted enthusiastically. . . . the enthusiasm was reserved, for few people within the industry gave *F&M* more than a fighting chance."[61]

Charles Roberts's fling had grown into a hobby, and now the hobby had become an "industry." But the industry insiders were wrong when they predicted an early demise for *Fire & Movement*. Despite some inevitable unevenness in the quality of its reviews, *F&M* became a tremendous influence on gamers and designers alike. Reviews and designer responses were supplemented by "Forum" articles, which allowed designers, critics, and just plain wargamers to present a variety of viewpoints about many new or potential design innovations or any other issue of interest to the hobby. The magazine did indeed go through some difficult financial times, changing publishers and editors several times. Through all its travails, however, it survives to this day because it always met the hobby's burning need for the kind of free and open forum of ideas and opinions it provided.

Throughout the 1970s, wargaming and wargamers were maturing. The serious minded had begun to question themselves and their motives for playing games dealing with death and destruction and enjoying them so much. In the Dunnigan tradition, they were also questioning the games, their designs, and their designers. Some designers objected to such questioning, and especially to the depredations of some of the *F&M* reviewers. The company magazines like *The General* and *Moves* had allowed the designers and developers of Avalon Hill and SPI games, respectively, to wax poetic about the brilliance and innovation of their design concepts. Despite some amount of self-deprecation and self-criticism that such magazines (especially *Moves*) engaged in, there was no real "public" interplay between designers and consumers. The growing numbers of reviews in non–house organs, especially those in *F&M*, soon found the staffs of major game companies scrambling to present dissenting opinions. Perhaps surprisingly, many failed to grasp the fundamentally subjective nature of game criticism, bemoaning a lack of "objectivity" on the part of reviewers.

One early example of the dismay with which some game companies received negative reviews was provided by Donald Greenwood, editor of *The General* and developer of the game *The Russian Campaign*. In response to a negative review of that game written by Richard DeBaun and Frank Aker that appeared in *Fire & Movement*, Greenwood wrote: "I find no fault with being critical; I'm the first to condemn goodie-goodie reviews which do no more than list components. A critic who is afraid to criticize is not worth the name. But there is no denying that it has become highly fashionable of late for wargamers to point out supposed flaws and mock professional game designs by their own self-ordained criteria. There is nothing wrong with this *provided those criteria are shared by the majority of the hobby as well.* I contend however that Msr's DeBaun and Aker were so preoccupied with reaching the height of fashion that they violated the most basic grounds of objectivity and thus *are not qualified to stand in judgment for the hobby as a whole.*"[62] [Emphasis added.]

In some ways Greenwood was right. Some aspects of the review of *The Russian Campaign* were inaccurate, and others perhaps overly biased. Yet the longing for hobby-wide standards of

136 criticism and qualifications for reviewers (indeed, Greenwood later goes on to suggest a sort of numerical "Critic Credibility Factor" to rate the opinions of reviewers relative to the hobby in general) reflect too much of an academic approach to game criticism on the part of game designers.

In the sometimes more, sometimes less, adversarial atmosphere between critics and designers, an important fact was being overlooked. Not only were designs being subjected to greater public scrutiny, but designers other than those of Avalon Hill and SPI were also getting a chance to articulate their ideas to a wider audience. The cross-pollenization of design concepts and approaches that resulted from the increasingly open exchange and discussion led to a blossoming of new approaches and an explosion of new games.

INNOVATION AND EVOLUTION

Partially initiated by Jim Dunnigan and SPI, and lovingly fostered by the professional hobby press, the late 1970s and early 1980s were a period of tremendous dynamism for hobby wargaming. Origins, a national gaming convention, began under the sponsorship of Avalon Hill in 1975. It gave new focus and identity to the hobby, and served as a place for new games and designers to introduce themselves to the players face to face. New topics and new approaches abounded, and a new game without some innovation to advance the "state of the art" had a difficult time in the marketplace. Several important trends began to sort themselves out during this period.

Perhaps the most dramatic development was the appearance of the so-called "monster game." Monster games were beasts of enormous size and, often, equally enormous complexity. They arose in some ways from the belief that more detail meant greater realism. They were the ultimate embodiment of an attempt to push the scope of a game as broadly and as deeply as possible, with little regard for practical notions of playability.

The original Charles Roberts games had maps that measured 22" × 28", and the number of playing pieces seldom exceeded 100 or so. Even Avalon Hill's largest game of the 1960s, *Blitzkrieg*, was barely a third larger in map area and still had fewer than 400 counters. The "standard" SPI game weighed in at no more than a 24" × 36" map and 200 to 400 counters. Although larger "homemade" games were constructed out of paper **137** and shirt-cardboard by many hobbyists (including your intrepid author, much to the dismay of his mother), the professionally produced big game did not appear until 1973, when Marc Miller, Frank Chadwick, and Paul Richard Banner teamed up to create *Drang Nach Osten* (loosely translated as "thrust to the east") and form Game Designers Workshop.

Drang Nach Osten, or *DNO* as it is affectionately known among aficionados, was a wargamer's fantasy come true. It contained five 21" × 27" mapboards detailing much of eastern Poland, Hungary, and Rumania, and the western Soviet Union. Its 1,700 counters represented the Soviet, German, and Axis-Allied armed forces that fought on the Soviet front in 1941–1942. GDW also published an expansion kit (!) known as *Unentschieden* (*UNT*), containing an additional four half-maps and 1,900 more counters to allow players to re-create the rest of the Soviet-German war. As large as *DNO/UNT* was, it was only the tip of the iceberg for GDW. Their ultimate intention was to produce a series of games at the same scale, which could be linked and played together to cover the entire European Theater of Operations in World War II on a map covering some thirteen square feet.

Not to be outdone, SPI soon published a rival game, using four of its standard maps and over 2,000 counters. *War in the East* was only the forerunner of a series of SPI monster games. It was followed, predictably, by *War In the West* and a combined *War in Europe*. They even went so far as to produce a *War in the Pacific*, with seven maps and 1,600 counters. (Although your author survived multi-player games of *War in the East* and *War in Europe*, the Pacific contest eluded me. So too did the ultimate, the combination of *War in Europe* and *War in the Pacific*. Gaming folklore has it that some intrepid soul actually undertook to play this super-monster solitaire! It seems hard to believe, though, even of a wargamer.)

The initial wave of monster games at the scale of theater-level conflict was followed by the introduction of the "grand-tactical" monster. The first board game to deal with a single land battle in an area as large as that required by miniature games was Martial Enterprises' *Bataille de la Moskowa*, a re-creation of Napoleon's Battle of Borodino with roughly the same physical size as *Drang Nach Osten*.

138   Both *Drang Nach Osten* and *War in the East* were successful with gamers as well as critics because, though physically large, their mechanics were not overly intricate and difficult to master. Experienced gamers (and even inexperienced ones not put off by the sheer size of the beast) could pick up the basics of play quickly. This allowed the player to concentrate his attention on the broad sweep of strategy or the nitty-gritty of deploying and maneuvering units without having to consult the game every few minutes.

The big games also offered ready-made opportunities for multi-player gaming experiences. Playing in teams brought players into firsthand contact with problems in command and dealing with subordinates that smaller games often had to abstract into dreaded "idiocy" rules. Some personal anecdotes might shed the most light on how this could work.

When I was in graduate school, I let my friend Bob Shore talk me into a multi-player game of *War in the East*. There were to be six of us involved, three per side. The only rule was that each side had to complete its movement in half an hour. I was given command of the central sector of the Russian line, whose initial dispositions had already been established by the overall Soviet commander. Although a good player, this commander was sometimes not as meticulous as I would have liked, especially when faced with a long front and several hundreds of units to deploy. Unfortunately, the commander of the German forces opposite my sector was both meticulous and observant.

As the war opened with the German assault, I watched in horror as several panzer divisions overran elements of my front line and plunged deep into the rear, encircling a large portion of my forces while the German infantry pounded on them in the front. With the best positions for a secondary defensive line already in German hands and many of my reserve armored formations cut off, I faced my first movement phase.

I also faced a very personal lesson in the meaning of the euphemistic gaming term "panic." As the clock ticked away I struggled to grasp the extent of disaster and to find ways to counter it. All hope of making careful and precise calculations of how many combat factors I would need to place in a given hex to prevent German overruns during the next turn proved vain. I scrambled in desperation to find some way of shoring up the

front. I was only partially successful, but I was completely de-
moralized. In little over one hellish hour, I had had more firsthand experience of the frustration of having to deal with a failure
by a higher command and the panic induced by a surprising and
lightning-like penetration of defenses that had looked strong than
I would ever care to repeat.

Just so you don't think that I was a complete dolt as a player,
and incidentally, to point out another aspect of the effects of big
games on modeling reality, take the case of a game of *Terrible
Swift Sword* (the Battle of Gettysburg) in which I was also involved as a graduate student. In this case, I came into the game
somewhat late, after the Confederate Army had pushed the Union
forces back during the first day's fighting. I was given command of
some battered and disorganized Confederate divisions on the
right center of the army. These units were obviously in no condition to make a major push anytime soon, and the Union player
opposite me was all but ignoring them.

Slowly and patiently I regrouped my scattered brigades and
posted them within extreme striking range of the Union line. All
the while I collected as many artillery batteries as I could steal
from the corps reserve. Still the Union did little to strengthen its
defenses or take advantage of the Confederate disarray to counterattack. My opponent had clearly written off this sector of the
battle line as a quiet zone and was spending little or no time on
it. That was to prove a fatal mistake.

I had seen the flaw in the Union dispositions. After nearly
three hours and some five or six game turns of seemingly doing
nothing more than shuffling counters around, I struck. My artillery swung into action and my infantry charged to the assault. I
had the sweet pleasure of seeing the same sort of surprised and
panicked look on my opponent's face as I'm sure mine held during the *War in the East* game. He had allowed himself to be
lulled into a false sense of security and had responded by spending too much of his attention on other sectors of the front. The
result was precisely the kind of surprising and successful attack
that so often happened historically under those conditions.

The monster games that were capable of producing such experiences were certainly impressive, and not the least impressive
element of their existence was the fact that they were actually
played. But getting involved in that kind of effort was not some-

140 thing all players wanted to do, and few if any wanted to do all the time. A balance was clearly needed, and that balance was found in the creation of what came to be called the mini-game.

There were two basic types of mini-games. The first, represented by SPI's *Stonewall*, used the full, complex system of its monster-game progenitor (in this case *Terrible Swift Sword*), but put it into a smaller physical package and a more manageable situation. In this way was born a series of games collectively known as *The Great Battles of the American Civil War*. Once they learned the basics involved in such a game series, players could readily adapt to any new game in that series. The trend toward series games has continued into the 1980s, with Avalon Hill's *Squad Leader* and *Advanced Squad Leader* games, Game Designers Workshop's *Third World War* series, and Victory Games' *Gulf Strike/Aegean Strike* and *Sixth Fleet, 2nd Fleet,* and *7th Fleet* games and West End Games' Civil War series (*South Mountain, Shiloh, Chickamauga*).

The second type of mini-game was, in a sense, a ready-made series game, at least as initially conceived in SPI's Quad games. Quad games were a set of four small games packaged together (although individual elements could be purchased separately). Quad games, like *Blue and Gray* or *Napoleon's Last Battles* were specifically designed to create interesting games using small maps (18″ × 22″) and relatively few counters (100 or less). In keeping with their limited size, they also usually had fairly short rules (eight to twelve pages) and fairly fast playing times (one to two hours).

In many ways, the Quad games and other small, simple games, were designed in the hope that they could play a role in introducing new players into a hobby that, to some, seemed increasingly dominated by the overly complicated and unplayable games of which the monsters were but the extreme example. The need for simple introductory games was reinforced by another major trend, which can be characterized as a push to include ever greater volumes of "hard data" in the games.

Some of the most vocal proponents of this trend were all but obsessed with technical minutiae. Drawing in some ways on the same emotional sources as the miniature-gamer's emphasis on precise detail in the painting of uniforms and vehicles, the push for increased volumes of data was most prevalent among enthusiasts of tactical games, especially those dealing with World War II armored combat. Avalon Hill's *Tobruk* carried the trend to an extreme.

Tobruk was designed for Avalon Hill by Harold "Hal" Hock, a card-carrying systems analyst. The game dealt with some of the series of desert battles fought between the Germans and British in the vicinity of Tobruk during Rommel's 1942 Gazala Offensive. Game counters represented individual tanks and guns and infantry squads or weapons teams. The "map" was simply a field of empty desert-tan hexes, each of which represented some 75 meters across.

Hock's underlying motivation for his design was a "desire to simulate every significant *event* as finely as possible."[63] Each weapon was thus rated for the number of rounds it could fire (at either a new target or a previously acquired one) in a game turn, and also for the probability of hit and damage against each possible target as a function of range. Furthermore, fire was resolved on a round-for-round basis. The result was a game dominated by tables and die-rolls (up to 200 and more in an average game). It was also a game players either loved or hated.

Hock had spent years modeling the firepower effectiveness of the various weapons in the game, using ballistic measurements and operations research techniques. He described his approach as "a careful reduction of all available data from technical and historical sources into easily understood systems of play which could be used to expose the data to players of the game." His goal was "to include every possible influence on the effectiveness of weapons and personnel either directly or by assumption into mathematical models built into the game." It was perhaps a noble goal and certainly one in keeping with the systems analysis philosophy. It was also unachievable.

Tobruk was technically superb and praised by those gamers who thirsted for technical detail. But it was also damned by many. Its modeling of the technical capabilities of systems was hard to criticize, but detailed technical accuracy and gaming realism are not synonymous. The precision of *Tobruk*'s weapons-effects model was offset by a lack of sufficient consideration for environmental effects such as haze, smoke, and dust. To make matters worse, the game, at least in its original incarnation, gave players few suggestions about the proper tactical doctrine for

142 fighting the types of battles it depicted. Too often this lack of guidance resulted in players' attempting to engage at extreme ranges and rolling dice interminably for little or no tangible gain. The resulting word-of-mouth did much to damage the game's reputation and limit its popularity. Even reasonably favorable reviews like the one headlining *F&M #*1 were unable to propel the game to general popularity.[64]

Fortunately, the *Tobruk* approach and the push for more and more technical data were not the only influences on the hobby's expanding attempts to quantify key elements of warfare. Because games rested ultimately on mathematical representations of reality, designers had to find ways of quantifying those aspects they believed to be important, even if they were the classic "unquantifiables" of leadership and morale.

Once again, Dunnigan and his cohorts took the lead in efforts to "quantify the unquantifiable." In games like *Leipzig* and *La Grande Armée*, Dunnigan and Young began to rate the capabilities of combat commanders. Dunnigan also introduced the concepts of command control and panic in games like *Panzer Armee Afrika* and *American Civil War*. Indeed, one of the major contributions of the large "grand-tactical" games like *Terrible Swift Sword* was their incorporation of troop quality and morale, concepts long popular in miniatures gaming, into the board-gaming world.

Similar advances were made in the treatment of broader concepts. Logistics, the perennial stepchild of wargamers interested only in fighting battles, became more important as designers tried to represent its effects more accurately. Limitations on a unit's or army's ammunition supplies were used in tactical games like *Terrible Swift Sword* and *Tobruk*. Even the broad scope of supplying pre-modern armies with food and fodder was modeled brilliantly in *1812* and *The Crusades*. Finally, a full-scale treatment of the different supply classes needed by modern armies was incorporated into GDW's *Operation Crusader*.

Perhaps even more fundamental, designers began to be more willing to experiment with different ways of representing the location and movements of military forces. The simple square or hexagonal grid that virtually defined board wargaming was joined by more and more "area-movement" systems, seen in classics like *Diplomacy*, and point-to-point movement systems in

Army-Navy maneuver, Newport, Rhode Island, November 1887. The North Atlantic Squadron under Rear Admiral Luce passes Fort Adams at the entrance of Narragansett Bay. (*Harpers Weekly* illustration courtesy of the U.S. Naval War College Museum)

One of the earliest illustrations of wargaming at the Naval War College. *Harper's Weekly* illustration of 1895. (Courtesy of the U.S. Naval War College Museum)

An apparent fleet-level action in Luce Hall at the Naval War College, about 1914. (Courtesy of the U.S. Naval War College Museum)

Wargaming at the Naval War College, Pringle Hall, circa 1947. (Courtesy of the U.S. Naval War College Museum)

A Pringle Hall Board Maneuver, about 1947, showing some of the early postwar gaming equipment. (Note the large-scale model ships in the background.) (Courtesy of the U. S. Naval War College Museum)

A logistics wargame in Sims Hall of the Naval War College, 1952. The logistics course began in 1950, headed by Rear Admiral Henry E. Eccles. (Courtesy of the U.S. Naval War College Museum)

Another Pringle Hall game, in 1949, showing the use of "talkers" (the commander third from the right) and transparent status boards (center). (Courtesy of the U.S. Naval War College Museum)

Manual wargaming nears the end of its heyday at Newport in this 1955 Pringle Hall game, showing one of the screens used to deny players information about their opponent's formations. (Courtesy of the U.S. Naval War College Museum)

The Status Board of the Naval Electronic Warfare Simulator (NEWS) in Sims Hall of the Naval War College, 1957. (Courtesy of the U.S. Naval War College Museum)

The game floor of the WARS, Sims Hall, 1981. (Courtesy of the U.S. Naval War College Museum)

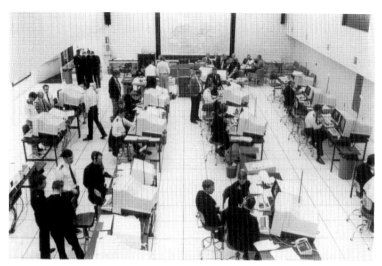

The game floor of the Naval Warfare Gaming System, Sims Hall, 1985. (Courtesy of the U.S. Naval War College Museum)

The Origins '87 National Gaming Convention, Baltimore Convention Center, 4 July 1987. Gamers playing tactical board games using cardboard counters (foreground) and lead miniatures.

War at Sea, introductory naval board game dealing with the Atlantic theater of World War II. (Courtesy of The Avalon Hill Game Company)

The popular board game *Victory in the Pacific*, companion game to *War at Sea*, but a more challenging treatment of the Pacific theater of World War II. (Courtesy of The Avalon Hill Game Company)

Naval War, a card-game introduction to wargaming. (Courtesy of The Avalon Hill Game Company)

The Origins '87 National Gaming Convention, Baltimore Convention Center, 4 July 1987. The dealers floor, showing just part of the crowd of hobby gamers evaluating the latest game releases.

The U.S. Air Force's Wargaming Center, Maxwell Air Force Base, Alabama. (Courtesy of the U.S. Air Force)

The game floor at the Air Force Wargaming Center during a global-level game. (Courtesy of the U.S. Air Force)

Planning the air campaign over NATO's Central Front during a Maxwell wargame. (Courtesy of the U.S. Air Force)

High-level planning session during a game at the Air Force Wargaming Center. (Courtesy of the U.S. Air Force)

The French Army of a hobby Napoleon deploys to attack the British reverse-slope position during a miniatures game in 1989. Note the protractor, rule, and hexagonal-shaped terrain sections, basic equipment for the miniatures gamer of the 1980s. (Courtesy of Larry Bond)

PACIFIC WAR

The Struggle Against Japan
1941-1945

From the American Civil War to World War II, Korea, and beyond.
Modern boardgames allow players to explore many of the most significant military campaigns in history (both past and "future"). (Courtesy
of Victory Games, Inc.)

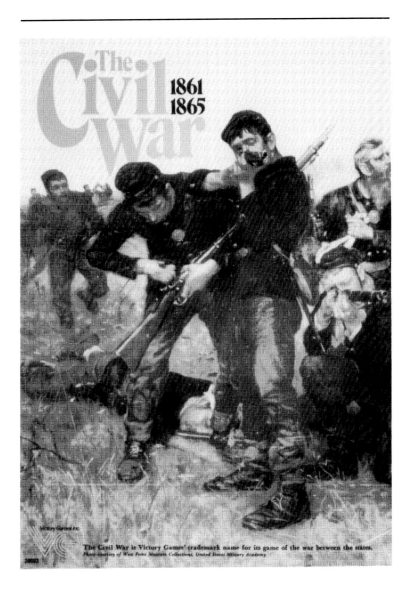

The Civil War is Victory Games' trademark name for its game of the war between the states.
Photo courtesy of West Point Museum Collections, United States Military Academy.

THE
Korean
War

JUNE 1950 –
MAY 1951

Victory Games Inc.
VICTORY GAMES, INC., New York, NY 10001
©1986, Victory Games, Inc., NY, NY 10001. Printed in USA.
THE KOREAN WAR is Victory Games, Inc., trademark for its
operation level game of the Korean War.

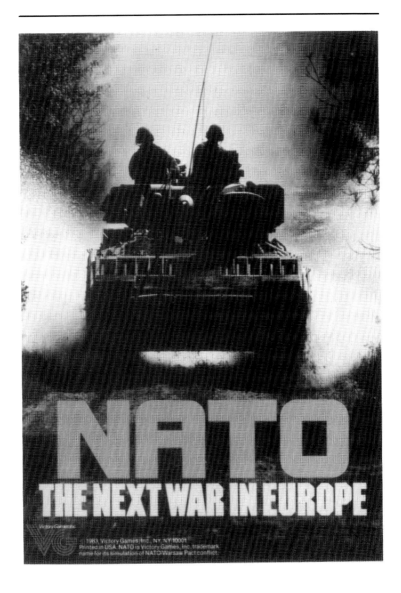

1983, Victory Games, Inc., NY, NY 10001.
Printed in USA. NATO is Victory Games, Inc. trademark
name for its simulation of NATO/Warsaw Pact conflict.

The 1980s have seen a wealth of boardgames published for the naval enthusiast. These are among the most popular. (Courtesy of Victory Games, Inc.)

Land, Air and Sea Combat in the Persian Gulf

GULF STRIKE

2nd EDITION

UPDATED
1988
VERSION

15

VICTORY GAMES, INC., New York, NY 10001
Division of Monarch Avalon Inc., Baltimore, MD 21214
1983 Victory Games, Inc., N.Y., N.Y. 10001
Printed in USA
GULF STRIKE is a Victory Games, Inc. trademark for its simulation game of contemporary conflict in the Persian Gulf.

SIXTH FLEET

MODERN NAVAL COMBAT IN THE MEDITERRANEAN

Victory Games Inc.

© 1985. Victory Games, Inc. NY, NY 10001 Printed in USA
SIXTH FLEET is Victory Games, Inc., trademark for its
simulation game of modern naval warfare in the Mediterranean.

2nd FLEET

MODERN NAVAL COMBAT
IN THE NORTH ATLANTIC

© 1986 Victory Games Inc., NY, NY 10001 · Printed in USA
2nd FLEET is a Victory Games, Inc. trademark for its
simulation game of modern naval warfare in the North Atlantic

games ranging from the early and little-known *Confrontation* to **143**
Napoleon and other games.

More radically, the traditional sequential system of play in
which first one player and then the other moved and then attacked, was more and more frequently replaced by more intricate
systems. All of the new approaches strived to achieve a more realistic feeling of simultaneity and action/reaction.

The old Avalon Hill games of *Bismarck* and *Midway* had
introduced simultaneous and hidden movement to the hobby by
means of duplicate playing boards separated by a screen. Players
searched for the enemy by calling out various locations, relying
on an honor system to provide each other with the appropriate
information. Such an approach proved difficult to adapt successfully to the operational conditions, the larger map size, and the
greater number of pieces in a ground combat game. Instead, designers began to employ purely simultaneous systems that required players to plot the activities of their units for each turn.
Such systems appeared in games like *Manassas* (an amateur
game later purchased by GDW); Dunnigan's *Kampfpanzer, Desert War*, and *Mech War '77*; and Battleline's air combat game
Air Force.

These plot-based systems never were very popular because of
the great amount of paperwork they required. Other games, like
Tobruk and *October War* compromised some simultaneity by
allowing players to alternate moving or firing one unit at a time.
This approach did away with much paperwork and seemed more
acceptable to most players.

Many more intricate systems were devised, including those
that integrated movement and combat in an attempt to reflect the
costs of battle in time (for example, in Dunnigan's monster game
of a NATO/Warsaw Pact conflict, *Next War*). No longer would
one player be free to conduct all his operations in a turn before
the opponent could react. An array of techniques far too numerous to mention here was introduced to allow for sudden shifts in
initiative and much more integrated play.

These and other innovations in game techniques had been
spearheaded by Dunnigan and SPI. As hobby board gaming entered the 1980s, however, that influential and inspiring source of
new ideas was destined to stay behind. In the winter of 1981–
1982, the hobby press was full of the biggest news in decades.

144 Simulations Publications Incorporated had ceased operations under the Dunnigan-Simonsen team, and its assets had been assumed by TSR Incorporated, best known for their (gasp!) fantasy-role-playing system *Dungeons and Dragons.*

HOBBY GAMING ENTERS THE EIGHTIES

Although TSR took over the name and logo of SPI, it could not sustain the creative spirit that Dunnigan and company had breathed into the hobby. After a short hiatus, *Strategy & Tactics* resumed publication. *Moves* disappeared as a separate entity, but some of its features were incorporated in the new *S&T.* Some of the TSR/*S&T* games were very well done and received favorable reviews in the hobby press. John Prados's *Monty's D-Day,* about the battle of the British/Canadian beaches at Normandy, was perhaps the most successful.

TSR also attempted to continue producing boxed games under the SPI label. Some of these games had already been in the old SPI's mill when the walls came crashing down. *A Gleam of Bayonets* was a large game of the *Great Battles of the American Civil War* series, which dealt with the battle of Sharpsburg/Antietam and was hailed for advancing the quality of that system significantly. *Battle Over Britain* was another of these long-awaited games, with an interesting and uniquely successful system for simulating the air combat over Britain in 1940. TSR also produced revised or simply repackaged editions of old SPI games and a few new titles as well.

Despite TSR's best efforts, however, it was clear that the giant was dead. The bipolar hobby world, dominated by Avalon Hill and SPI, with a few lesser lights like Game Designer's Workshop and West End Games, and a host of smaller third-world companies, was gone. Yet the spirit of James Dunnigan had not died. Instead, it seemed to split in two and separate itself, part of it staying on the East Coast and part migrating to the west.

Dunnigan's eastern reincarnation took the form of Victory Games, a company formed by many of the old SPI team, and headed by Mark Herman. Herman had been a developer under Dunnigan, gluing together one of the last of the great monster games, *Next War.* Under his leadership, Victory Games formed as a subsidiary of the same larger conglomerate that owned Avalon Hill. But while Avalon Hill remained true to its philosophy of trying to produce middle-of-the road, finely crafted, "player's **145** games," Victory was more innovative and produced high-quality, boxed games with greater appeal to the fans of complex and intricate "simulations." In some ways, Victory Games played Pontiac to Avalon Hill's Cadillac.

Victory's initial batch of four releases took the hobby by storm in 1983. *The Civil War* featured a brilliantly conceived and imaginatively executed system of initiative and command control to produce what is arguably the most comprehensive and well-received game on that popular subject yet. In *Gulf Strike*, Herman designed the first game to integrate completely air, sea, and ground movement and combat in a game dealing with the Iran/Iraq war and the potential for superpower involvement in the Persian Gulf. *Ambush*, dealing with combat in Western Europe after Normandy at the man-to-man level, was a breakthrough in solitaire-wargame design. Even *Hell's Highway*, the only poor seller among the four, took a unique and innovative view of the infamous Arnhem campaign of *A Bridge Too Far* fame. In one fell swoop, Victory Games had become a force to be reckoned with in the hobby, and the philosophical heir to the Dunnigan tradition of innovation and elegant game-system design.

The second half of the Dunnigan tradition, the willingness to explore obscure topics and to produce a large number of "paperback" games, was taken over by Briton-turned-Californian Keith Poulter. Poulter made his mark in Britain as editor and publisher of *The Wargamer*, a *Strategy & Tactics*-like magazine with a game in each issue, as well as a combination of historical and gaming-oriented articles.

In the early 1980s Poulter brought his organization, known as World Wide Wargamers and, inevitably, 3W, to the United States and the little town of Cambria, California. He wasted little time in making a play to capture *Strategy & Tactic's* former role as the dominant magazine in the gaming hobby. Poulter started with the traditional bimonthly magazine, but quickly promised a shift to a monthly schedule. Skeptics were numerous, pointing out the difficulties of producing not only a magazine but also a game on even a bimonthly schedule, as well as the problems involved in finding and testing that many reasonably designed games in a year. Many also quaked at the high price a subscription would inevitably cost.

146   But Poulter was undeterred. Starting slowly at first, he eventually was successful in producing not twelve but thirteen games in a single year. In the process he also achieved the remarkable feat of improving not only the quality of the physical components of the games, but their design and development as well.

The number of games published by *The Wargamer* allowed 3W to publish successful games on popular topics (such as *Sturm Nach Osten* on the Russo-German front of World War II) as well as the occasional foray into the truly obscure (such as games dealing with Britain's eighteenth century Afghan wars, the fifteenth century siege of Malta, and the rise of Saudi Arabia).

Ultimately, Poulter's efforts at *The Wargamer* succeeded in wresting the Charles Roberts award for Best Professional Boardgaming Magazine away from *Fire & Movement* at both the 1986 and 1987 Origins conventions. Then, in the kind of flurry of activity so well-known of him, Poulter acquired the rights to *Strategy & Tactics* magazine from TSR, divested himself of the *The Wargamer* (now stripped of its game-per-issue format), and masterminded a merger with Diverse Talents, Inc., the most recent publisher of *Fire & Movement*. By 1988, Poulter had thus become the controlling force behind both *Strategy & Tactics* and *Fire & Movement* magazines, truly names to conjure with in the wargaming hobby.

In early 1989, however, Poulter surprised the hobby world yet again by selling *Fire & Movement* to Christopher Cummins, the new owner of *The Wargamer*. Poulter planned to begin publishing a new computer-game-oriented magazine and to expand his operation into the family-game market. Furthermore, the merger with Diverse Talents had brought with it another wargaming magazine, *Battleplan*. *Battleplan* was oriented toward gameplay topics, such as variants and replays of existing games as well as hints on strategy and tactics. Poulter's new team gave increased attention to *Battleplan* and its relationship to *Strategy & Tactics*, in an apparent attempt to reestablish the old *S&T*/*Moves* tandem masterminded by Dunnigan.

Although perhaps not as prolific or innovative as Jim Dunnigan, Keith Poulter is a successful game designer in his own right, especially of Napoleonic and American Civil War titles. His ultimate influence on the hobby is one of the intriguing questions of the coming decade.

PROFESSIONAL GAMING REAWAKENS

While hobby gaming was literally exploding with new game-design concepts, new ideas for using old tools, and new tools altogether, professional wargaming, though not completely stagnant, was advancing far more slowly. As the war in Vietnam wound down and belief in the utility of nuclear weapons as the means to prevent all future conflict dissipated, defense analysts and decision makers began to address the difficult questions posed by the needs of limited, or even global, conventional war. Wargaming's popularity as a research tool staged a comeback.

In the new breed of professional wargame, an increased emphasis on speed of processing went hand-in-hand with an increased emphasis on detailed physical accuracy. The availability of large computers, like the one used in the NWGS, seemed to hold the promise of resolving the ancient tension between realism and playability by the crude expedient of brute force. And brute force was all that was available to the majority of mathematicians, operations researchers, and computer programmers who were tasked with designing wargaming systems. Lacking the enormous and varied game-design experience of the hobbyists, they sought the wrong solutions, often to the wrong problems.

The big systems sometimes succeeded—the Navy's WEP-TAC system, for example—but more often than not they promised much more than they were capable of delivering. To supplement their capabilities, professional wargamers were forced to adopt the types of ad hoc combinations of human judgment and manual and personal-computer-based techniques that came into use at the Naval War College.

Despite the defense community's continued problems with design, implementation, and execution of playable and realistic games, gaming and gaming techniques became increasingly important at nearly all levels and among all the branches of the military. By 1977, the disarray and sense of frustration in the community of analysts and wargamers was "mildly akin to the atmosphere among theologians on the eve of the Reformation."[65] In Leesburg, Virginia, in September of that year, the Office of Naval Research sponsored a conference titled "Theater-Level Gaming and Analysis Workshop for Force Planning." The conference brought together many of the leaders of the professional

148 wargaming and analytical community. Included among their number were Andrew W. Marshall and none other than James F. Dunnigan.

Marshall was, and still is, the director of Net Assessment in the office of the secretary of defense, responsible for making overall evaluations of the relative military strengths of the United States and the Soviet Union. Marshall had searched in vain for existing analytical models and wargames that could be of use to him. It was time for something new.[66] Three years later, Marshall would issue a request for proposals to build a new methodology for conducting high-level strategic and policy analysis. The request would eventually result in the development of the Rand Strategy Assessment System (RSAS) (originally known as the Rand Strategy Assessment Center, or RSAC).

Marshall's search for new approaches, and his recommendations that the professionals should look more closely at what the hobbyists were doing, eventually prompted Dunnigan to schedule an appointment to visit him in the Pentagon.[67] Dunnigan told Marshall that he and SPI could produce a global-war simulation for less than one-fifth of the cost of Marshall's initial exploratory contracts with the larger professional organizations. The result of the meeting was a contract for Dunnigan to develop such a system for wargaming on the global level, a system that became known as the Strategic Analysis Simulation, or SAS. Although Dunnigan left SPI soon after the project got underway, Mark Herman took over development responsibility and finished the initial version of the system. SAS is still in use, particularly at the National Defense University where it "allows the players to analyze decisions at the National Command Authority, Joint Chiefs of Staff, Commander in Chief and fleet commander levels. It simulates armed conflict between major land, sea and air forces, supports a major National Defense University study exercise each year, and is refined by students through the National Defense University electives program."[68]

At the same time that Marshall began his search for RSAS and commissioned Dunnigan to produce SAS, the Army War College at Carlisle, Pennsylvania, was reintroduced to wargaming by Dr. Raymond M. Macedonia, a former army officer, who "with the unpaid assistance of Jim Dunnigan and some of his hobby-shop board games, started developing a new military center for gaming and model-making."[69]

The biggest push for wargaming at Carlisle was soon provided by Army Chief of Staff General Edward C. Meyer. Meyer urged Macedonia, as only an army four-star can urge, to develop an easily set-up and easily used system to help the army conduct contingency planning. The task fell to Fred McClintic, one of the programmers who were helping Macedonia computerize some of Dunnigan's old games. McClintic, "using new principles of computer architecture, created what became known as the MTM—the McClintic Theater Model."[70]

The model was designed with two purposes in mind. The first was the traditional wargame role in the U.S. Army, that of training. The second was to "provide the basis for evaluating and modifying corps commander's strategy and tactics in the Tactical Command Readiness Program."[71] Its earliest applications, however, were even more ambitious.

The first application of this model was for the Chief of Staff of the Army's [CSA] Contingency Planning Seminar in November, 1980. At that time the CSA, DCSOPS [Deputy Chief of Staff for Operations] and 11 of 16 active division commanders used the model as a tool to examine alternative courses of action. In April, 1981, the model was used by VII Corps commander and all of his division and brigade commanders plus their senior staff officers in Exercise "Cold Reason" to walk through their general defense plan. In July, 1981, the model was used by REDCOM/RDF [Readiness Command/Rapid Deployment Force] staff officers to examine sustainability/logistics and operational problems and opportunities for exploiting enemy weaknesses. In addition to these applications, it was used by the Strategic Studies Institute for the Parametric Force Analysis Study (PFAS) examination of the effects of the Geneva Protocols in the NATO theater of operations, and in several Army War College Advanced Courses."[72]

To accomplish its many and varied purposes, the McClintic model was designed to allow the players themselves to operate their computer terminals, giving commands to their forces in a free-form style using certain key words. Showing the influence of commercial board games, the model "is based on a variable-size hexagonal grid network and is applicable to any part of the world."[73]

The model was also designed to allow play using only two

150 terminals, although in that case the game controller would have to share a terminal with one of the players. Although compatible with computerized graphics hardware, the original system did not require graphics capability to play. Instead, players relied completely on a standard map overlaid with a clear hexagonal grid.

One of the key, and in many ways revolutionary, features of MTM is its use of continuous time to drive the game, as opposed to the more frequently seen Red/Blue team sequencing. Unlike the "event-driven" protocol used in many strictly analytical computer simulations, "MTM is time driven; that is, battle time proceeds at a predetermined rate faster than real time regardless of when the next event occurs. This gives wargame commanders a chance to change or countermand orders up to the time that order is actually executed."[74] In this adaptation of the "game clock" idea, used in the Naval War College's NEWS and NWGS systems, to a game based on a relatively small, transportable computer, the McClintic Theater Model may have scored one of the professionals' few successes in beating the hobby gamers to the punch with a new innovation.

THE PERSONAL-COMPUTER EXPLOSION

Sometimes a revolution begins with a "shot heard 'round the world." Sometimes one begins with an obscure and long-forgotten article in a small magazine. Such were the origins of the electronic revolution in hobby wargaming.

The first published description of an electronic version of a hobby wargame seems to have appeared in the Nov.-Dec. 1974 edition of the Avalon Hill *General*. Interestingly enough, this article described a device that could serve as "an electronic mapboard which duplicates the map of the North Sea that is given in the standard *Jutland* game."[75] Thus, Jim Dunnigan's first professionally published design, *Jutland*, is accorded the status of serving as the basis for the hobby's first electronic game.

Simply described, the device was a wooden box on the opposite faces of which were inscribed the North Sea map used in the search phases of the standard *Jutland* game. In the center of each of the map's hexes was a metal rod, "connected, by wire, to another rod in the same sea square [hex] on the other side of the board. . . . The positions of ships are marked by alligator clips which are placed on the rod of the sea square in which that ship or group of ships is supposed to be. . . . Movement takes place just as it does in the standard *Jutland* game. The players move their clips from sea square to sea square and when both players have ships in the same sea square a light flashes which signifies contact."[76]

From such humble, basement workshop beginnings, the fuse was lit that, in less than a decade, would ignite an explosion of computerized wargames for the hobbyist. When personal computers began to become accessible to the hobbyist in the late 1970s, wargaming was not far behind.

The early personal computers, like the original Apple kits and the early TRS-80 and others, provided the wargamer with a means of storing and rapidly manipulating the quantities of mathematical data needed to play the games of the period. (Some of the earliest uses of computers revolved around automating the calculation of combat odds and determining engagement outcomes.) Not surprisingly, Jim Dunnigan was one of the early pioneers in the use of computers in "game assistance programs" and as an aid to game design. SPI had computerized some aspects of its business operations in the early 1970s, and in 1975 they began using the corporate computer to assist with game development— for example, testing the model of the Soviet economy for *War in the East*. An attempt in 1974 and 1975 to publish a newsletter dedicated to promoting the use of computers for wargaming proved premature, but by 1979, *Moves* magazine raised the issue again and advertised for the submission of game assistance programs for SPI games.[77]

Of course, it was not long before the fledgling game-assistance programs gave way to full-scale wargames for the computer. The early, first-generation games were often little more than military-style "arcade" games, similar to those designed for home video-gaming systems. (Today, many commentators on computer wargames give these early games, like Instant Software's *Ball Turret Gunner* and The Cornsoft Group's *Missile Attack*, short shrift or ignore them completely. Yet such games were the forerunners of important later developments like Microprose's *Silent Service* and deserve at least a passing mention.)[78]

The second-generation games followed rapidly on the heels of their arcade-style brethren. These games, typified by Avalon Hill's early *Midway Campaign* and *North Atlantic Convoy Raider*,

152 were basically simplified electronic derivations of board wargames. Strategic Simulations, Inc., better known as SSI, began
its rise to success with this type of game, one of the earliest of
which was *Computer Bismarck*. A reviewer at the time described
the game in the following terms. "Currently it is to the hobby
what the original *Bismarck* by Avalon Hill was in its first release
back in the sixties. We still have a lot to look forward to, but you
can enjoy it now."[79]

The second-generation games typically retained much of
the conventions of board wargaming; the hexagon-style maps and
sequenced player turns were perhaps the most prominent. But
the early designers, such as Gary Grigsby, creator of *Guadalcanal
Campaign* and *War in Russia*, quickly saw the advantages of the
computer's "potential for artificial intelligence. The first use they
saw for it was to command the enemy forces, thus offering even
the most isolated wargamer an intelligent opponent who is always
ready to play. Since many wargamers play solitaire much of the
time, a built-in opponent has inestimable value. This feature
is so important that virtually all computer wargames offer some
solitaire capability."[80]

As the computer-wargame designers gained experience,
however, they cautiously began to deviate from their board-game
roots. The obvious uses of the computer (to introduce limited intelligence and more complex combat resolution systems) began
to be supplemented by adapting the computer's artificial intelligence to have it "act as the player's subordinates. Like a real commander, the player issues general orders, and intermediate commanders then try to carry them out."[81] In addition, the shift
to continuous-time games, similar in philosophy to the system
underlying the McClintic Theater Model, began to make itself apparent by the mid-1980s. Games such as SSI's *Combat
Leader* and Microprose's *Crusade in Europe* were forerunners of
a third generation of computer wargames. The critical links to
that third generation, however, were Microprose's *Silent Service*
and Strategic Studies Group's (SSG) *Carriers at War*.

Silent Service, as its name implies, deals with the activities
of the U.S. submarine force in World War II. Unlike Battleline/
Avalon Hill's board game *Submarine*, however, *Silent Service* literally places the player on the bridge or in the conning tower of
his boat as he plans and carries out an antishipping mission in the
Pacific.

Silent Service is a "flight simulator" for submarines. The **153** player literally sees what his real-life counterpart would see, as animated ships move across his field of view. Or, he can descend to the sonar room to watch the display of acoustic contacts on the screen. To the arcade-game tradition of visual appeal and the need for quick mental and physical reactions, *Silent Service* added accurate systems simulations and historical scenarios to create one of the most intense, personally involving wargames yet invented. It was the forerunner of Microprose's newer and even more impressive *Red Storm Rising* game, based on the best-selling novel by Tom Clancy (and, incidentally, hobby-game designer Larry Bond). The *Red Storm Rising* game simulates the operations of U.S. nuclear submarines in a hypothetical future war.

At a higher level of command, SSG's *Carriers at War* also scored a major breakthrough. CAW, as it is known, deals with the major clashes of aircraft-carrier fleets in the Pacific War. Designed by Australia's Ian Trout and Roger Keating, CAW allows a single player to "assume the role of a (lowly?) task group commander and let other players, or the computer, handle other functions. Or, a player can be the theater military commander and delegate subordinate functions to other players, or to the computer. Beyond the solitaire and multi-player options, design-your-own scenarios allow: altering terrain, choosing from 7 types; up to 63 different airplane types in up to 127 squadrons, for over 4,000 total aircraft; up to 24 land bases; 63 ship classes and 48 task groups, as many as 32 carriers plus 215 other ships from transports and destroyers through seaplane tenders and submarines, to full battleships such as the *Yamato* and the (original) *New Jersey*."[82] Using a menu-driven system to help the player carry out his game functions, and relying on a continuous-time system to simulate the action, "*Carriers at War*, quite simply has *no* peer on the public market today. One of the reasons that it has become one of today's 'giants' is that it stands squarely on the shoulders of several of yesterday's highly-rated board and computer games."[83]

SSG's success with *Carriers at War* in 1984 was followed by the publication of *Europe Ablaze*, a strategic/operational game of aerial warfare in World War II. Then they proceeded into the field of land combat, producing *Battlefront*, which emphasizes the limitations on the player-as-commander's ability to move and direct the activities of each of his subordinate formations. In just

154 a few short years, this Australian company came from down
under to lead the way into the long-awaited third generation of
computer wargaming, games that dispense with the structure of a
board-gaming mentality and board-gaming conventions. This
successful penetration of a non-U.S. company into a field dominated by Americans is, perhaps, symbolic of the slow but steady
progress of postwar wargaming outside the United States.

WARGAMING OUTSIDE THE U.S.

The importance of American experience in wargaming
since 1945 is hard to overstate. As was the case with so many
other things in the postwar world, the U.S. had become the
dominant force in defense analysis and wargaming among the
nations of the free world. Yet it would be remiss to end this chapter without discussing the perspectives of a few other nations with
wargaming experience and programs.

The defeat and destruction of German and Japanese military power had also destroyed their wargaming traditions and
"corporate memory." Both nations were infected with the same
brand of confusion between wargaming and operations research
that afflicted the U.S. In West Germany, for example, "the term
'Gaming' is considered to subsume all techniques that employ
some kind of model of competitive interaction between actors
who may be individuals or groups. . . . They encompass the entire range, from *training games* in which the trainee acts out the
life role for which he or she is being trained, to *analytical games*
which describe the interactions in terms of highly abstract objective functions."[84] Figure 6 shows the relationship between reproducibility and degree of abstraction on one hand and the resource requirements and operational realism on the other for the
various things considered parts of the gaming category in Germany. (Note that "Analytic Games" are equivalent to what we in
the United States would simply call analysis.) "All of these types
of games are presently being used in one form or another in the
German defense establishment for purposes of training as well as
planning and analysis support at all levels, i.e., from weapon system design up to strategic contingency analysis."[85]

Indeed, most European nations, including many analysts
in the United Kingdom, view wargaming in much the same
light. There are exceptions, however, to this commingling of

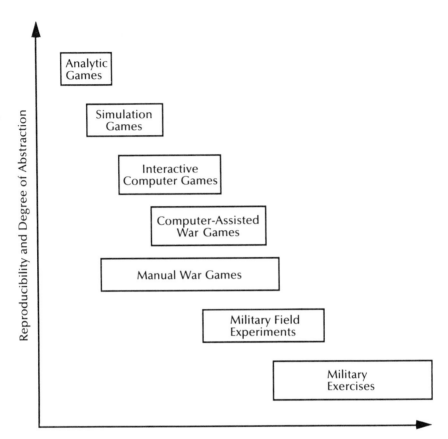

Figure 6. A West German view of analysis, wargames, and military experiments and exercises. (Adapted from the talk *Gaming and Simulation—A German Perspective* presented by Reiner Huber at the symposium *Serious Games for Serious Questions*)

wargaming and other techniques. One British analyst, in a paper dealing with "mathematical models and computer simulations," distinguishes them from wargames, which "fall into a completely different category by virtue of the strong element of human decision-making and the interactions among players. They deserve discussion in their own right."[86] Even more to the point, "Canada has persisted over the years in manual war games and shunned the mathematical battle models."[87] The Japanese, too, have only recently made the decision to reenter the field of real

156 wargaming in a serious way, contracting with U.S. firms to produce a system similar to that used at the U.S. Naval War College.

The Israelis, however, have made extensive use of wargaming in almost every aspect of their society.[88] Israeli military gaming is based on the notion espoused by Sun Tzu that military power is composed of human beings, weapons, and wisdom, of which wisdom is the dominant force. To the Israelis, gaming is a central tool in the development of wisdom.

Israeli games are used to sensitize people and organizations to potential situations and so help avoid surprises. They are also used in the traditional roles of planning and training aids. Games have been used with great success to identify the types of intelligence information that might be required in certain foreseeable situations, and to assess and evaluate possible threats and identify vulnerabilities.

The basic principles of Israeli military gaming include:

- Play with real people (usually subordinates as close to the primary decision maker as possible).
- Play with real data.
- Play in real time (crucial to avoid teaching wrong lessons).
- Play according to real manuals, orders, and procedures.
- Play real problems.
- Play in real psychological environments, putting players into the field to create as many realistic pressures as possible.

Perhaps you have noticed a single word common to all these principles. The Israeli stress on realism is so strong as to be almost obsessive.

Perhaps most interestingly, the Israelis believe strongly in playing games with multiple groups, often in parallel, seeking different viewpoints and alternative solutions to the problems presented. This philosophy extends at times even to "Red" play. Typically, the opposition is provided by the usual source of opposition players, the intelligence community. Sometimes, however, a second opposition team may be put together consisting of the "outsiders" in the intelligence community, those whose opinions differ from the conventional wisdom.

Not surprisingly, perhaps, the Soviets are also heavily involved in gaming and simulation. Although to date there is little information available in the West about Soviet wargaming techniques, some things are known based on information provided by defectors and emigrées from the USSR and Afghanistan.[89]

The Soviets define the study of war in extremely scientific terms, terms significantly different from the less-formalized view of the U.S. "Sometimes the Soviets use wargames to analyze, test, and develop ideas that emerged during training exercises. Or they use games to develop theories imposed by policy decisions."[90] A Soviet wargame is initiated as a result of a directive from the headquarters ordering the game.

The critical element in organizing the game is the establishment of the control group, which in turn organizes the umpire group and player teams. The control team thoroughly prepares the materials required for supporting the game and prepares the location in which the game will be played. The control team also prepares the "concept of the operation" for running the game and designs the "initial situation" or scenario.

The control team also decides on the procedures that will be used to regulate player actions and presentations of results. The chief controller specifies the role of the umpires, and the control team also determines the time intervals for action. The Soviets view "each game as a vignette focused on specific critical actions or activities which the control wants to investigate or to teach to the players. The method for handling the decisions made by both sides . . . is one of the most critical aspects of the game design. In many if not most games there is only one time step. The next time step becomes another wargame."[91]

Interestingly, "the plan for the wargame includes the *method for drawing conclusions* from these team decisions."[92] The control team also prepares the plans for conducting the critique of the game, which occurs immediately after play is complete.

"The Soviets use wargames for research and development . . . [and] for developing actual war plans. There are examples of instances in which a wargame proved a plan was faulty and they changed the plans."[93] Although the original Soviet term for wargame was literally translatable as "war game," their new term corresponds more closely to "staff training exercise." "Wargames are a favored method for training command and staff personnel . . . [and they] consider wargames so important that officer participants have been promoted or demoted on the basis of their performance in wargames."[94]

Of course, the Soviets also recognize the shortcomings of wargaming. "One of the most difficult aspects of wargames is the creation of the proper psychological conditions."[95] Wargames are

158 certainly not a panacea. "The Soviets conduct much research and development prior to the wargame and incorporate the results in turn in further development work. The process is continuous as more questions are answered and they in turn raise other questions. Therefore wargames are important, but only a part of a larger system."[96] In many ways, however, there is a rigidness in the larger system, which uses wargames too often to study the standard "prototype" situations of Soviet theory, each of which will typically already have an accepted solution, at least in general terms. Whether the Soviets can thus benefit from the competitive drive of wargaming is unclear.

In addition to the types of professional wargaming activity described above, hobby wargaming has also developed outside the U.S., but again on a lesser scale. The strong tradition of miniatures gaming in the U.K. was never supplemented by a strong board-gaming tradition. Although many U.S. games were imported, only a very few indigenous U.K. games were ever produced. Indeed, one of the most productive U.K. wargame companies was 3W under Keith Poulter, who moved his business to the more dynamic and lucrative U.S. market.

Hobby wargames have also been published in France, Sweden, and especially Italy. In recent years, Japanese board-game publishers like Hobby Japan have produced first-class products, both of domestic designs (*Pacific Fleet*) and of games by U.S. designers (*Norway, 1940*).

It is in Australia, however, that non-U.S. hobby wargaming seems to be most prolific and most interesting. Led by John Edward's pioneering company, Jedco, several Australian concerns have made significant contributions to gaming in the 1980s. The Australian Design Group produced the large and popular game *World Aflame*, dealing with the entirety of World War II, following an equally impressive effort representing the Napoleonic Wars (*Empires in Arms*). Panther Games entered the fray with an extremely innovative design dealing with, what else, the Russo-German war (*Trial of Strength*) and led the way into new realms of computer wargaming with *Fire Brigade*. As discussed earlier, however, when it comes to Australian computer-game companies, Strategic Studies Group, producers of *Carriers at War*, *Europe Ablaze*, *Battlefront*, and many others, clearly dominates the market.

NEW DIRECTIONS OR OLD DISAPPOINTMENTS?

As has been the case with so many aspects of the modern world, the progress of wargaming in the past 25 years has far outstripped the developments of the previous 2,500. Even so, wargaming has been unable to escape from the recurring cycle of popularity and disfavor that has dogged its steps throughout its long and checkered history.

In the professional world, the status and importance of the wargame as both a tool and a discipline seem unquestioned. By the early 1980s the need for some sort of automated planning and decision aids to help the military commander sort through an increasingly complex modern environment was a major factor in the establishment of the JCS-sponsored Modern Aids to Planning Program (MAPP). One of the principal elements of the MAPP is a complex computerized simulation and wargaming system known as the Joint Theater-Level Simulation (JTLS), an evolutionary development of the concepts underlying the McClintic Theater Model. In 1987, the Senate Armed Services Committee held closed, secret hearings on the practice of high-level wargaming and its role in the formulation of national military strategy and policy. Led by the Naval War College, the principal military service schools are all involved in wargaming to a greater or lesser extent.

In the Pentagon, the joint chiefs of staff continue their long-standing program of political-military gaming to explore potential crisis situations around the globe. Even President Reagan's Strategic Defense Initiative has come under the scrutiny of academic, industrial, and military wargames. The director of the Office of Net Assessment in the Defense Department felt so strongly about the potential value of wargaming in assisting the policy formulation process at the highest levels that he has contracted for the creation not only of a large, "automated wargaming" system (the RSAS), but has also arranged for the development of a series of smaller, more traditional games relying on human players, drawing designers from the wargaming hobby into direct involvement with the Department of Defense.

From its small and tentative beginnings, the wargaming hobby has grown in numbers and stature, as hobby gamers bring their background and experience to bear on real-world problems.

160 The hobby games themselves cover the broad spectrum of military history, science fiction, and, perhaps most importantly, current and near-term military affairs. From "traditional" leaden miniatures, to "classic" paper and cardboard, to the most advanced "high-tech" personal computers, hobby wargames attempt to exploit the best of the old and, at the same time, experiment with the best of the new.

Despite all the new ideas and new techniques evident in both worlds, however, wargaming today stands on the threshold of another downturn in popularity and acceptance, both with the hobbyists and the professionals. Hobbyists have made it difficult for new players to join their ranks because many of their games have become so large and complex that they are almost impossible for an inexperienced player to grasp. Professionals still struggle with unwieldy or unworkable systems or ideas and continue to make promises they can neither retract nor keep. To prevent the type of downturn wargaming has experienced in the past, it is necessary for those who believe in the value of wargames and wargaming to draw on the experiences of that past to identify the central principles of wargaming, and to define the sphere of its legitimate application. Part II addresses these issues.

PART II

PRINCIPLES

4

THE NATURE
OF WARGAMES

Over the course of their long history, wargames have taken many different forms and served many different functions. From the early topographic maps and detailed charts of the rigid *Kriegsspiel* or the lead soldiers and spring-loaded cannon of *Little Wars*, to the small personal computer of *Carriers at War* or the large mainframe computer of the Naval Warfare Gaming System, the tools and techniques continue to evolve. Despite the obvious differences in form, the substance of what makes all these different devices wargames remains the same. Yet in its modern usage, especially, the term *wargaming* has been defined in many ways, and it is important to examine what wargames are and what they are not.

In its broadest application, the term *wargame* is used to describe any type of warfare modeling, including simulation, campaign and systems analysis, and military exercises. In the 1979 edition of Webster's New Collegiate Dictionary, for example, "war game" (note, two words!) is defined as a "simulated battle or campaign to test military concepts and uses. Conducted in conferences by officers acting as the opposing staffs," or "a two-sided umpired training maneuver with actual elements of the armed forces participating." Such definitions contribute to imprecise

164 and sometimes misleading discussions of the subject. In fact, these broad definitions are especially a problem among professional wargamers and other members of the defense community, who too often look to wargames for solutions to problems they cannot usefully address.

What wargaming is not is usually even less obvious than what it is. First and foremost, wargaming is not analysis, at least not in the defense community's usual sense. It is not a technique for producing a rigorous, quantitative or logical dissection of a problem or for defining precise measures of effectiveness by which to compare alternative solutions.

Nor is wargaming real. Despite the similarities of gaming language and the gaming experience to important aspects of actual military operations, its abstractions are many, and too often they are not obvious to those without real-life experience.

A wargame is not duplicable. You cannot replay a wargame changing only the random numbers (or the die rolls). The chances that two independent games will produce the same sequence of decisions and outcomes are so low as to be negligible.

A more restricted and more useful definition is that a wargame is a warfare model or simulation whose operation does not involve the activities of actual military forces, and whose sequence of events affects and is, in turn, affected by the decisions made by players representing the opposing sides. In the end, a wargame is an exercise in human interaction, and the interplay of human decisions and the simulated outcomes of those decisions makes it impossible for two games to be the same. As a result of all those factors, wargaming is not a panacea for learning about or solving the problems of warfare. Its forte is the exploration of the role and potential effects of human decisions; other tools are better suited to the investigation of other more technical aspects of reality.

ELEMENTS OF A WARGAME

The focus on human decision making implies that a good wargame must be structured to help human players make decisions and to allow them to learn about the effects of those decisions. Although different games have taken drastically different structural approaches, they all have made use of the same key elements. Whether it is a hobby game designed to help two people

study an historical campaign during the course of an afternoon, **165**
or a week-long professional game designed to help hundreds
of people explore the potential for future warfare, a good war-
game needs:
- Objectives
- A scenario
- A data base
- Models
- Rules
- Players

and, especially for professional games,
- Analysis.

Every wargame has objectives. The most abstract hobby
game may limit itself to providing its players with an enjoyable
afternoon's entertainment in a quasi-military setting. The most
detailed professional game may strive to give its players the most
realistic training experience possible short of the use of actual
hardware and live ammunition. Whatever those objectives may
be, the more clearly they are stated, and the more judiciously
chosen, the more likely that the game will be a success.

Well-defined objectives are essential to a professional game.
In specifying the objectives, game sponsors, designers, and analysts must clearly identify how and in what ways the game can
provide the type of experience and information needed to achieve
them. The statement of objectives should be as specific as possible to allow game-design efforts to focus on those elements critical to the production of that experience and information, and to
the assimilation of training lessons or the collection of research
data. A wargame's objectives should be the principal drivers of its
entire structure.

The scenario sets the stage for the game by placing players in
specific situations and giving them a context for their decision
making. The scenario can have a significant, if not overwhelming, effect on the decisions players are able to make. As a result,
the game designer must carefully determine how the scenario
may affect the factors he is most interested in exploring. Detailed
scenario descriptions should allow the players to understand
those factors and how they arose, so that they can understand
how the underlying assumptions may affect the scope of their decision making.

166   The data base contains the information players may use to help them make decisions. Typically, this information includes the forces available, some measure of their capabilities, physical or environmental conditions, and other technical facts. Because of its importance to decision making, the data base must present clearly and concisely the information players would reasonably have available to them in an actual situation, and it must do so in a manner easy for them to use during play.

A set of models, usually a combination of look-up tables and mathematical expressions, translates the game's data and the players' decisions into game events. Models must be flexible enough to deal with unforeseen player decisions. They should be designed to allow the data base to change without requiring major changes to the models themselves. Their mechanisms should reflect accurately those factors most important to the decision-making levels represented by the players. As much as practicable, the question of whether a model will depend on random numbers should be driven by the underlying real-life process. Just as real battles are affected by chance, game battles should usually reflect the role of luck in carrying out any operation.

But models alone are not enough. A wargame must also have a set of rules or procedures that dictate how and when to apply the models. These rules and procedures help sequence game events, and allow for accurate chains of cause and effect, or action and reaction. They must also ensure that the players receive the appropriate quantity and quality of information during play. Where possible, the rules should try to introduce errors into the player's information and delays into the execution of orders or arrival of new information. Such procedures, designed to simulate the "fog of war" are often the most difficult to devise. In most professional games, and in a growing number of hobby games, one or more umpires may be employed to monitor player actions and control the flow of information. As faster, more powerful, and more flexible computers become available, even to hobbyists, many of these umpire functions are being assigned to the machine.

Ultimately, however, there is one function that cannot be given to a machine without fundamentally changing the character of a game and turning it into something else. A real wargame must have human players whose decisions affect and are affected by the flow of game events. A game is most effective when those players can be cast in operational roles and be given the information and responsibility required to make the decisions appropriate to those roles.

In professional wargames, a seventh critical element must be added to those described above. This final element is game analysis. If the objectives of the game define the information that must be extracted from its play, game analysis assures its capture. In a training game, analysis will usually consist of an instructor's observation and critique of the student's play. In a research game, analysis focuses on understanding why decisions were made. A good analysis plan, outlining where observers should be placed and what they should be looking for, is essential, but the process of game analysis is not simply one of mechanics or even observation. The data collected during game play is only the raw material for the synthesis of insights and identification of issues.

Even hobby games are subjected to analysis, and two types of such analysis are evident in the hobby press. The more common type is the game review or critical analysis of the game system. Good reviewers focus on the designer's viewpoint of reality and on his approach to representing that viewpoint through the game system. Reviewers assess the strengths and weaknesses of the viewpoint and of the resulting game mechanics and evaluate how well the game plays. The second type of game analysis concentrates on an actual play of the game. This type of analysis, often seen in Avalon Hill's *General* magazine under the title "Series Replay," consists of a turn-by-turn recounting of play in which each of the players summarizes the considerations underlying his or her decisions, and a neutral commentator evaluates those decisions on the basis of prior experience with the game and full knowledge of both players' plans. Although such analysis seldom attempts to project beyond the game board into reality, it is otherwise somewhat similar to the type of analysis done for professional games.

Each of the elements of a wargame will be discussed in greater detail in the chapters dealing with game design, development, play, and analysis. Before getting into such details, however, it is useful to understand some of the more arcane language of wargaming. We begin with the broadest characterization of what a game is all about.

168 *THE LEVEL OF A WARGAME*

Hobby Games

The publishers of hobby wargames often advertise their products with terms like "tactical-level simulation" of a particular battle, or "operational-level game" of an important campaign, or "strategic study" of an entire war. Some games, such as the large-scale re-creation of the battle of Gettysburg known as *Terrible Swift Sword*, are even christened with the high-sounding title "grand tactical game." Although there is some amount of marketing hype involved in this terminology, the distinctions among levels of game are real and important.

The two most popular criteria for categorizing the level of a game are based on the command level represented by the player, or on the level of activity at which a player's decisions are implemented and evaluated. The words used to describe these levels are usually the same words used to define the major aspects of the military art: strategic, tactical, and operational.

A player's decision level is strategic if his responsibility extends to allocating resources, possibly including economic and political resources as well as military forces, to fight and win an entire war. A player is making tactical-level decisions if he is most concerned about positioning relatively small numbers of men and weapons to apply violence directly to the enemy; that is, to fight battles. The operational level game is less easily described; here the player is concerned with maneuvering relatively large forces so that they can be positioned to win the battles they fight, and so that those battles can help win the war. In the sense of decision making, then, the level of the game reflects the scope of the players' decisions.

In the sense of evaluation of results, on the other hand, the level of the game most closely represents the degree to which the processes of military violence are abstracted. In a tactical game, the emphasis is on that process itself, not only on what the outcome is but also on the physical processes by which it takes place. An operational-level game may abstract this evaluation process, considering only the types, numbers, and relative positions of forces involved in combat and ignoring the lower-level details of who fired what weapon at whom and when. A strategic game might abstract this even further, resolving actions over large expanses of territory and time by comparing gross amounts of forces involved and ignoring completely the niceties of maneuver and application of fire.

A classic example of a tactical game is the popular modern-naval miniatures system of *Harpoon*. In this game the players command one ship or a small group, possibly with aircraft in support. Their goals are usually fairly limited and for the most part involve engaging and destroying an enemy force while avoiding the same fate. Individual missiles, bombs, or guns are fired at specific targets and detailed physical effects are calculated (such as whether a round disables a particular weapon or sensor system).

The board game *2nd Fleet* is a good example of an operational-level game. The players are concerned with an entire theater (in this case the Norwegian Sea) and large numbers of forces of many types. The action takes place over a span of days or weeks as opposed to *Harpoon's* hours or days, and the objectives of the players involve control of wide expanses of ocean, not merely the destruction or avoidance of specific enemy units.

An example of a strategic game is *Pacific Fleet*, dealing with the entirety of World War II in the Pacific. In this game, players represent the supreme commanders of all the forces involved in the war, maneuvering entire fleets and armies. The action extends over months and years, with individual engagements resolved simply by comparing the effective strength of the opposing forces in a large expanse of ocean or ground territory.

Hybrid Levels

It is possible for games to combine a higher level of decision making with a lower level of combat resolution. Strategic-operational and operational-tactical games are often popular with the hobbyist. For example, *Pacific War* covers much of the same strategic ground as *Pacific Fleet*, but includes a much more operationally oriented combat system. *Carrier Battles* deals with the air-sea campaigns around the Solomon Islands in World War II at an essentially operational level, including detailed tactical rules for resolving combat.

The principal difficulty involved in such hybrid games is their potential for allowing the player to fall between two stools. For example, a player who concentrates on his high-level command position to the detriment of his low-level tactical skills may

170 find himself fighting the right battles but losing them because he is not as able as his opponent in designing defensive ship screens. On the other hand, a good tactical player may never lose a battle, but could lose the game for the lack of a coherent campaign plan. This disconnect between the level at which the player is required to perform his principal decision making function and the level at which those decisions are resolved can be even more pronounced in professional games. Fortunately, many of the problems arising from the disconnect can be solved when a sufficient number of players are available to allow a division of command responsibility along the lines of the different command levels.

Professional Games

Professional games are seldom characterized on the basis of the level of combat resolution. Instead, the most useful categorization of modern professional wargames combines the idea of geographic scope with the level of the decision making. This scheme defines three broad classes: global/strategic, theater/operational, and local/tactical.

Global/Strategic

In global/strategic games, the primary decision makers represent opposing National Command Authorities. Typically, the goals of such games are to improve the perspective of the participants, test strategies, and identify important issues at the global level. Early examples of these games focused attention principally on pre-hostilities and the politics and force deployments involved in the transition from peace to war, the D-day engagements, and questions regarding escalation or war termination. Recent games have expanded their explorations into the capacity of the opposing sides to sustain prolonged conventional war, and into the effects of threats to the strategic nuclear balance within the context of conventional war. The primary output of these high-level games is qualitative, consisting typically of game narratives with some interpretations of events and little numerical data. Games on this scale usually require large numbers of people and great amounts of time to play, and are seldom, if ever, repeated under identical conditions.

The Global War Game series conducted at the Naval War College for the past ten years has had a growing effect on the development of strategic thinking in the navy and in the defense community as a whole. The scope of the effort, both in the global nature of its scenarios and the numbers and diversity of its participants, has served as a catalyst for raising important research issues. It has facilitated the exchange of ideas among professionals who seldom have the chance to interact (such as experts in Soviet political affairs and scientists working on advanced technological concepts).

Theater/Operational

The primary decision makers in theater/operational games are typically cast as commanders in chief of the unified or specified commands in a particular region. Some games actually combine multiple theaters to achieve a pseudo-global scope, but because decisions are made at the theater level, these games are closer to the operational rather than strategic scale.

Theater/operational games are usually designed to explore specific issues and identify strategic, operational, and tactical problems in the theater. Often they point out areas in need of further study. Such games focus on the force levels and employment options necessary or feasible for carrying out specific military missions. Although the output of these games is similar in nature to that of global/strategic games, there is a tendency to have the same group of players run through the game more than once and also to generate more numerical data.

Theater/operational games may be the level of game most usefully and most frequently employed for many research areas. They are used to "preplay" or test plans, from exercise designs to fleet war plans. When well designed, such games force participants to deal with the same situations they might face in an actual operation. They allow commanders and their staffs the chance to explore how and why their plans might be able to deal successfully with the problems they have perceived. They also provide fertile ground for identifying unforeseen difficulties and unexpected solutions.

Local/Tactical

The primary decision makers in this category of game are generally battle-group commanders or below. (In land-combat games, the equivalent command level is typically that of a corps or

172 divisional commander or lower.) As is the case with the global/ strategic games, a principal purpose of tactical games is to give their participants an improved perspective. Local/tactical games are also used to compare various tactics and force structures and, perhaps even more than in the other types of games, to identify topics for further analysis.

Typically, local/tactical games focus attention on force levels and tactical deployments, weapon and sensor performance, and interrelationships among various warfare areas. The outputs of these games usually have a greater balance of qualitative and quantitative results than do the others. The number of times players replay a local/tactical game varies, but it does tend to be higher than in either of the other two categories.

Games at the tactical level, dealing as they do with the most basic realities of warfare, are the most difficult to design "accurately." There is an unfortunate tendency to focus more closely on numerical "results" even when the reliability of such results does not deserve it. Yet, when properly designed and executed, tactical wargames can be incomparable tools for exploring the feasibility of tactics, identifying the hidden assumptions, both valid and invalid, on which such tactics might be based, and highlighting their potential strengths and weaknesses. The ideas that evolve from such games are often easily translated into concepts that can be further tested and refined by at-sea or field exercises.

Table 1 compares the three categories of professional games defined above.

OTHER WAYS OF CATEGORIZING WARGAMES

In addition to the level of the game, wargames can be characterized by the mode of evaluating activity, the number of players or "sides," the information limits imposed on the players, the game's style, and the game's instrumentality.

Mode of Evaluation

We have already discussed the notion of evaluating the outcome of player decisions in terms of the level of detail involved in the process. A more basic distinction, almost as old as wargaming itself, is that between "free" and "rigid" evaluation techniques.

Table 1. Levels of Professional Wargames

	Global/strategic	Theater/operational	Local/tactical
Primary decision maker	National Command Authority	Commanders in chief	Battle group or lower
Goals	Give participants a better perspective, test a strategy, identify key issues, facilitate exchange of ideas.	Explore specific issues; identify strategic, operational, and logistical problems in theater or exercise; identify areas for further study.	Give participants a better perspective, compare various tactics/forces, identify critical factors and areas for further study and exercise/testing.
Focus	Prehostilities and transition politics and force deployments, the D-day shootout, and escalation/war termination.	Necessary/feasible force levels and employment options for accomplishing specific military missions.	Force levels and tactical deployments, weapon and sensor performance, and interrelationships among warfare areas.
Primary output	Qualitative. Narratives and interpretations with little numerical data. Typically only a single game run.	Qualitative. Narratives and interpretations with some numerical data for more support. Typically a small number of games run.	Balance of qualitative and quantitative results. Number of iterations may vary, but tends to be higher than in other categories.

174   Free evaluation techniques rely principally on the capabilities and experience of an umpire to determine the outcomes of encounters without requiring the use of any particular set of data or mathematical models. Rigid evaluation techniques center around precisely those types of formal models seldom used in the free methods. Not surprisingly, most hobby games employ the free method, but increasingly a mixed approach is preferred in which the umpires use both their experience and some models to make their evaluations.

Number of Players

Most wargames today are two-sided. One player or team will represent each of the two major contending parties or coalitions. The two-player game has been a staple of hobby wargaming since before the invention of chess, and is equally popular in purely military professional games. In the latter, however, there is often a third, non-player team called "control," which handles matters outside the scope of the player decision levels, and carries out the umpire functions.

Multisided games, with three or more independent active player teams, are most often seen in the realm of diplomatic or political-military games. Indeed, one of the wargaming hobby's most celebrated games, *Diplomacy*, is designed for seven players representing the great powers of Europe at the turn of the twentieth century. In the professional gaming world, political-military games have been played by the Organization of the Joint Chiefs of Staff, the National Defense University, and other organizations, to allow the exploration of multilateral interactions in a wide variety of scenarios.

At the opposite extreme is the one-player game, in which a single player or team plays against a pre-programmed opponent, or in which the control team assumes direction of the opposition side in addition to its usual functions. One-player games are frequently employed for training purposes, the instructors controlling the play to reinforce specific lessons.

Most hobby gamers are familiar with one-player games in several different guises. Because of the complexity and time-consuming nature of most board wargames, opponents can be difficult to find, and many players play a large number of games solitaire. However, the solitaire game in which a single person plays both sides of a two-player game, is not what we mean by a **175** one-player game. The true one-player game, in which the opposition's activities are directed by the game system, umpire, or other devices, was a rarity in hobby gaming until just recently, when Victory Game's *Ambush* series and *Mosby's Raiders* game popularized two different approaches to games designed specifically for solitaire play. The advent of personal computers has made solitaire gaming more feasible because the computer can provide a pre-programmed opponent that is ready to play the game whenever the human player is. One-player games dominate the hobby "simulator" market, as evidenced by *Silent Service*, *Red Storm Rising*, and *F-19 Stealth*.

Information Limits

One of the most dramatic distinctions among wargames is that between an "open" game and a "closed" game. An open wargame allows all players essentially free access to all available information about each side's forces and capabilities (but not about plans!). Typically, such games use a single situation map on which forces from both sides are, for the most part, openly deployed. A closed game better simulates the "fog of war" by introducing limits on the information available to the players. Such games usually attempt to restrict player knowledge of their own and their enemy's forces to that which could realistically be expected from available sensors. Closed games almost always require some sort of computer assistance unless they are very small in size or scope. As a result, true closed games are a recent development, despite the long history of attempts to introduce some form of limited intelligence into games. The difficulties of successfully playing a closed game without computer support are strongly indicated by the very small number of such games published by the wargaming hobby.

Style

Another major categorization of wargames is their separation into seminar games and system games. In a seminar game, opposing players discuss the sequence of moves and countermoves they are likely to make in a given situation and agree on what interactions are likely to occur. A control team then assesses the results of those interactions and reports back to the players. This process is

176 repeated for each of the "moves" in the game. Seminar games often use moves of various lengths of real time (time steps) and so tend to resolve different periods of action at different levels of detail. Not surprisingly, seminar games tend to be open games, at least for the most part. They also tend to be limited to the professional gaming world where research, discussion, and learning are usually more important than the personal competition that plays a more significant role in hobby gaming.

Most hobby wargames are system games. A system game substitutes a structured set of highly detailed and specific rules and procedures for the more informal discussion process of the seminar game. Player decisions are implemented through the medium of the system, and actions that the system does not provide for may not be undertaken. Once the decisions are enacted, the system determines any interactions and outcomes. Although not all system games are closed, nearly all closed games must be system games.

In some ways system games are closely related to rigid umpiring, while seminar games are related to free umpiring. Just as professional gaming has endured a long-standing competition between rigid and free umpiring, there has been a similar competition between seminar and system games. The increasing advocacy of computer gaming has reinforced the pressure for increased use of system games. The continued lack of realistic operational data on which to build a detailed system argues for the utility of the give-and-take of a seminar game, where bad ideas or skewed evaluations are subjected to greater scrutiny. As with most such debates, the proponents of either side too often fail to see that each approach has merit.

Instrumentality

Although it may be possible to play some forms of wargames without the use of any prepared materials, most wargames require a set of tools to keep track of and to display data, force locations and movements, and interactions between opposing units. In chess, for example, there is the simple board of sixty-four squares that allows each of the thirty-two playing pieces to be positioned and moved. A typical hobby board game will make use of a mapboard on which is superimposed a hexagonal grid

containing hundreds of hexagons on which players maneuver **177**
some 200 or more cardboard counters.

The strictly manual games represented by the above examples were the norm of wargaming for over a hundred years.
The tools were relatively simple: maps, charts, notebooks of data
and orders of battle, perhaps a set of written rules and procedures.
As computers grew in capability after World War II, they began
to be adapted to wargaming, first by the professionals and then by
the hobbyists.

Computer-assisted games use machines ranging from desk-
top personal computers to very large mainframes. The machines
are used to help keep track of the positions of forces, their movement, weapons capabilities and other critical, data-intensive
pieces of information. The extensive bookkeeping required to
keep track of dozens of pieces of information about many hundreds of units virtually prohibited the types of detailed global wargames that are possible today from even being contemplated before computers became available.

The Rand Corporation has been in the forefront of an effort
to extend the role of the computer beyond that of capable assistant or sometime opponent. In their Strategy Assessment System, Rand is attempting to replace not only one but both human
players by computer software. Using artificial intelligence and
expert-system concepts, Rand is seeking a way of automating the
decision process to allow multiple replications of "wargames"
with changes in basic assumptions. Such computer-played games
are difficult to categorize as true wargames. They may develop
into a new, but related, tool, but their dangers are many. Early
enthusiasm for the potential of Rand's approach must be tem-
pered by the realization that no system of this sort can do more
than allow us to explore the implications of our own assumptions
as embodied in the computer programs. This is far different from
exploring the dynamic interaction of two human beings respond-
ing to ever-changing challenges.

SO WHAT IS WARGAMING GOOD FOR?

After all the discussion of what wargaming is and is not, the definition of a wargaming taxonomy and the dissection of a wargame's anatomy, where do we stand? What is all this stuff good

178 for, after all? Is it really the answer to every defense decision maker's dream, the oracle that allows us to peer into the future? Or is it just an elaborate toy, an incursion into the world of serious issues and serious decisions by something better left on the gaming tables of warmongering adult hobbyists who somehow never outgrew their destructive childhood impulses?

Aperiodically, the press of one country or another decries wargaming, either as a hobby or as a professional tool. One of the latest such incidents occurred in West Germany, where that country's "Examination Board For Publications Harmful to Young People" decided that hobby wargames were "morally corruptive and coarsening for the young user" and therefore their advertisement and sale should be prohibited. (Presumably, the older user was already corrupt beyond hope of redemption.)[1]

Sadly, there are wargaming hobbyists, and professionals, to whom such concerns and criticisms may well apply. They see only what they want to see, and learn from a game only those things that confirm their own preconceived notions. But such people are not unique to wargaming, and if wargames did not exist they would simply find other devices that could allow them to realize their dreams of conquest or their certainty in the depth of their understanding and the value of their own insight. For the vast majority of thoughtful, serious, and eager-to-learn gamers, however, the conclusion of H. G. Wells's book *Little Wars* vividly describes the fundamentally humbling experience that even the victorious wargamer reflects on in his quiet moments alone.

I have never yet met in little battle any military gentleman, any captain, major, colonel, general, or eminent commander, who did not presently get into difficulties and confusions among even the elementary rules of the Battle. You have only to play at Little Wars three or four times to realize just what a blundering thing Great War must be.

Great war is at present, I am convinced, not only the most expensive game in the universe, but it is a game out of all proportion. Not only are the masses of men and material and suffering and inconvenience too monstrously big for reason, but—the available heads we have for it, are too small. That, I think, is the most pacific realisation conceivable, and Little Wars brings you to it as nothing else but Great War can do.[2]

In today's world, where the critics of wargaming of both the professional and hobby variety castigate it for encouraging a fondness or longing for the illusory challenges and glories of war and therefore for fostering a readiness to engage in it, the insights of one of history's most famous, vocal, and sincere pacifists can be illuminating. Wells recognized that there are powerful emotional and intellectual challenges found in war that are difficult to discover in other forms of human activity. Rather than decry these challenges as abhorrent, Wells sought to reproduce as many of them as possible in a safe and sane way, on the game floor rather than on the battlefield. War, or Great War, as Wells characterized it, was and is abhorrent and unimaginably chaotic and destructive, especially for those who have never experienced it firsthand.

But wargaming, Wells's Little Wars, not only provided a harmless setting in which human beings could face some of war's challenges without destroying lives, property, or nature, but also taught something of the reality of Great War to those not familiar with its practice. For those in the military, wargaming (*Kriegsspiel* in Wells's lexicon) was an important tool for training and education, an ideal means for "waking up the imagination."[3] Rather than condemn the study of warfare by the military services (whose profession it was, after all), Wells the pacifist argued cogently that if "Great War is to be played at all, the better it is played the more humanely it will be done." He saw "no inconsistency in deploring the practice while perfecting the method."[4]

In today's world, where another Great War could well mean the end of mankind, wargaming is not just a hobby for the overeducated or a toy for the military-industrial complex, it is a way to help us understand the nature of the beast, and through that understanding to, if not tame it completely, at least prevent it from devouring us.

The Usefulness of Wargaming

In the end, the role of wargames of all types, sizes, and levels is to help human beings investigate the processes of combat, not to assist them in calculating the outcomes of those processes. Wargame designers, players, and analysts, as well as critics and decision makers who judge the validity of a game or define its results only in terms of what happened, not why, or only in terms of "lessons learned" not "issues raised," have lost sight of what a

180 wargame really is and where its main benefits are to be found. Wargames can help explore questions of strategy, human decision making, and war-fighting trends. They are of little use in providing rigorous, quantitative measures to "objectively" prove or disprove technical or tactical theories. Instead, they can often provide the kernel of new theories that can be tested with other tools.

Wargaming is most productive when used as an organizing and exploratory tool or as an explanatory device. It seems especially appropriate for exploring the dynamic nature of warfare. The design of the game (organizing) and the play and subsequent analysis of the game (exploring) form a circular chain or feedback loop, in which the questions and issues arising from one play of the game can reshape or reorganize the game system itself to make it a more accurate representation of reality.

As an organizing tool, wargaming helps designers and participants tie their thoughts together and give them a more operational focus. Designing a game requires comprehensive and coherent study and modeling of the interplay of different types of forces carrying out different kinds of missions for different sorts of reasons. The successful translation of quantitative and qualitative tactical analysis into a workable and meaningful game system requires a basic understanding of all possible force interactions, how and when they might occur, and what might determine their outcome. It also requires an understanding of how players interact as they develop different approaches to the problems posed by the game. Finally, it requires an ability to translate that understanding into intelligible and practical procedures so that the players can concentrate on making realistic decisions, not on remembering artificial rules.

When used as educational devices, wargames force the participants to begin translating what they have studied about strategy, tactics, or administration into something they can use in carrying out their mission or in understanding reality. Students at the navy's Surface Warfare Officers School may have been taught the rate of fire of a surface-to-air missile system, the reliability of the missile, and the radar horizon of a ship's search and guidance systems. They may have studied the speed and altitude of an enemy submarine-launched cruise missile and the number of such missiles a threat submarine might carry. They may even be aware that the time they might have available to react to an attack on their ship by such a missile submarine may be less than one minute. Yet, the true meaning and interconnections of all those facts are difficult to perceive in an abstract setting. By using a wargame to place students in "command" of a ship that is the target of such a missile attack, however, instructors can not only demonstrate the facts, but they can also allow the students to demonstrate their implications to themselves.

As an exploratory tool, wargaming can give players, analysts, and other observers and participants new insights, which can lead them to further investigation of the validity and the sources of their beliefs. Wargaming forces participants to look at reality from a different angle and can lead to fundamental changes in how they see that reality. If the initial design of a game incorporates well-known critical factors into its models and procedures, the play of the game and the questions and issues it raises can lead to the discovery of other factors whose importance may have been previously unsuspected or undervalued.

By explicitly allowing human decisions that are made under the press of time and on the basis of imperfect or incomplete information to influence the course of events, and by incorporating the capricious effects of randomness and "luck," wargaming comes closer than any other form of intellectual exercise to illuminating the dynamics of warfare. By illustrating the effect of these "unquantifiable" factors in concrete terms, wargaming also helps to illuminate the sources of that dynamism.

Finally, as an explanatory device, wargames can effectively communicate historical, operational, and analytical insights to hobbyists or to professional members of the defense community. The latest intelligence about threat operational doctrine or options can present commanders with new problems and challenge them to find feasible solutions. The operational implications of advanced weapon systems can be portrayed vividly by forcing players to deal with the opportunities and difficulties they present rather than by simply providing decision makers with numerical estimates of a limited number of technical parameters. The constraints of knowledge and capability under which historical commanders had to operate can be re-created to allow players and re-

182 searchers a fresh perspective on why events took place as they did, helping to offset the distortion and intellectual arrogance that too often accompany 20-20 hindsight.

Participants in wargames are not a passive audience. Their interaction with the scenario, the systems, and each other provides opportunities for the development of new insights. These insights can, in turn, prompt more detailed historical, operational, quantitative, and scientific analyses whose results can become incorporated in follow-on games. This process of sharing, testing, and revising knowledge and understanding is fundamental to the productive use of gaming.

Dangers of Wargaming

The power of a wargame to communicate and convince, however, can also be a potential source of danger. Wargames can be very effective at building a consensus on the importance of key ideas or factors in the minds of participants. They attempt to create the illusion of reality, and good games succeed. This illusion can be a powerful and sometimes insidious influence, especially on those who have limited operational experience. For example, a poorly designed game could allow players access to an unrealistic quantity and quality of information and so give those players a false picture of the worth of a weapon system that relies on just such unattainable information to be effective.

In wargames, as in any approach to study and analysis, there is always a possibility that intentional or unintentional advocacy of particular ideas or programs may falsely color the events and decisions made in a game and lead to self-fulfilling prophecies. The designer of a game has great power to inform or to manipulate. The players and others involved in the game and its analysis must be aware of this danger. They deserve and should demand an explanation of why events run counter to their expectations. They must be allowed, indeed encouraged, to be wary and skeptical and to question the validity of insights derived from the game until the source of those insights is adequately explained. If the reasons underlying an insight seem artificial, the insight may be a false one, and the game system may be in need of correction. These and other aspects of the roles and interactions of game designers, developers, players, and analysts are the subjects of the next four chapters.

5

DESIGNING WARGAMES

It is important to make one thing clear at the very start; designing **183**
a wargame is an art, not a science. Experienced military officers, practiced operations research analysts, and accomplished computer programmers are not necessarily capable of designing useful wargames. Although some or all of the knowledge and skills of such people are important tools for a wargame designer to possess, the nature of game design requires a unique blending of talents.

Wargaming is an act of communication. Designing a wargame is more akin to writing an historical novel than proving an algebraic theorem. The latter requires the use of deductive reasoning within a previously defined framework of knowledge; it proceeds from a set of assumptions to a set of conclusions by following the strict paths dictated by the rules of logic. The former requires the construction of a framework, the creative building of an internally complete and consistent world whose broad contours are contained within the bounds of its historical context. Yet, within those bounds many paths may exist, and not all of them may be apparent when the work begins.

So it is with a wargame; perhaps even to a greater degree

184 than with a novel. The wargame designer builds his world and provides a set of maps and instructions for exploring it. But it is the player, not the designer, who dictates which paths will be taken. Indeed, it is often the case that players will discover paths through the game world that even the designer did not perceive. As a result, and perhaps paradoxically, the end product of a wargame design must be much more structured, rigorous, and clearly delineated than any novel, precisely because the game's designer cannot dictate more than the broad limits of its player's explorations.

As an art form, all wargame design is based on some fundamental principles. These principles have been recognized in various guises from the earliest days of Helwig and von Reisswitz. Yet, the actual shape of any particular design will take on a unique form, a form that springs from the reasons for which the game was created, the nature of its participants and audience, and the talent and skill of its designer. Unlike music and literature, however, whose formal rules of pitch and grammar are codified and generally well understood (even by those who choose to violate them to achieve an artistic effect), game design has no real formalisms. Instead, it is dominated by individual style and by fashion, and in that respect is more like painting than other arts.

Previous chapters have described some of the changes in gaming fashion. The abstract and rigid formality of war-chess gave way to the representational realism and free umpiring of *Kriegsspiel*; the spring-loaded three-inch naval gun of *Little Wars* gave way to the detailed charts and tables of modern miniatures gaming; the gridded floor of the coffee mess at the Naval War College gave way to the computer graphics of the Naval Warfare Gaming System. Fashions change, and those changes result more often from the emergence of certain "landmark" game designs, created by designers with a new idea, than from the slow evolution of generally accepted design techniques.

The world of hobby wargaming clearly shows the influence of this phenomenon. Charles Roberts designed *Tactics* and launched the board-gaming hobby and industry. James Dunnigan designed *Tactical Game 3* and spawned an entirely new genre of board wargames. Roger Keating and Ian Trout designed *Carriers at War* and redefined the state of the art of computer wargaming. Each new ground-breaking design almost inevitably produced a flock

of imitators and a wave of innovations building on its fundamental ideas.

Since perhaps the mid-seventies, however, the design of
wargames for the defense community has suffered from the lack
of such revolutionary ardor. Most of today's professional wargames are based on long-established principles and techniques.
(One possible exception, the Rand Strategy Assessment System,
makes use of recent advances in artificial intelligence research,
but, as discussed earlier, it is not really a wargame.) The Naval
War College, which has done so much to maintain and advance
wargaming's popularity and usefulness, has nevertheless built
its reputation on a continuing refinement of the basic ideas
and methods espoused by Francis McHugh in the early 1960s.
The seminar wargame, albeit with increased computer assistance, remains the cornerstone of War College wargaming. The
Army's new JANUS system is little more than a variant of the old
McClintic Theater Model. Only OSD's attempts to introduce
"path games" seems to be breaking new ground.

Professional wargaming organizations (or, more accurately,
professional organizations involved in wargaming) are generally
built around a staff that spends most of its time adapting existing
game systems and facilities to the needs of next week's game.
They rely on many years of corporate experience and tend to rele-
gate the individual game designer to choosing which aspects of
the available tools will be used, not to designing new ones. In
many cases, they have succeeded quite well at producing useful
wargames. Unfortunately, the lack of innovation in technique
has resulted in stagnation. The older systems are beginning to
show their age, and fancy new facelifts like the Enhanced Naval
Warfare Gaming System or high-resolution color graphics will
not be able to reverse the trend.

The underlying cause of creeping obsolescence of game systems lies in the fundamental nature of game design itself. Ultimately, the goal of all wargame design is communication. Consciously or unconsciously, a wargame designer transmits specific
messages, concerns, and even conclusions to the players of the
game though the medium of its structure and procedures. As the
message changes, so must the game.

In a professional wargame, the source of such messages and
concerns is the game's sponsor. The sponsor is an individual or

186 organization who initiates the game and who wants the game to achieve certain goals. Usually the game designer is not the sponsor, and, as a result, the designer must first strive to understand the sponsor's intentions and objectives. He must then devise some effective means of transmitting those intentions to the players, thus achieving the game's objectives through their play.

The professional wargame designer must also remember that the game communication flows both ways, and he must structure the game's design to facilitate that flow. Through their participation, the game's players transmit their own concerns, questions, interpretations, and insights back to the sponsor. These elements of the wargame experiment are collected and reported, either directly through player comments, or indirectly through game analysis.

Hobby game designers are usually their own "sponsors." They communicate to the game's players their personal understanding and interpretations of the events their game portrays. There is, however, less of an immediate two-way exchange of information in hobby games. Hobby designers do get some feedback, though, either through direct contact with players, or through reviews and surveys conducted by hobby magazines devoted to criticism and analysis of games (and often of the historical situations on which they are based).

In both the hobby and professional worlds, the problems that designers or sponsors need or want to address change with the growth of new ideas or new technologies, or with the refinement of the designer's understanding and insight. The open, free market of the wargaming hobby shows the effects of this process of growth much more dramatically than the relatively closed and self-perpetuating society of professional wargaming and analysis.

To return to the example of James Dunnigan, his designs of *Tactical Game 3*, *Combat Command*, *Red Star/White Star*, *KampfPanzer*, and *Mech War '77* show the progressive development and growth of his ideas about how best to represent modern mechanized warfare in a game. In the professional sphere, only the Naval War College's evolving Global War Game series, largely under the managing inspiration of O. E. "Bud" Hay, exhibits a similar prolonged period of growth and development, but without quite the same sense of vibrancy as that exhibited by Dunnigan. Perhaps this shortage of real, revolutionary change that plagues the professional wargame is one of the sources of the up-and-down cycle of gaming's popularity.

As the tools of warfare continue to evolve at an ever-quickening pace, professional game designers need to break out of the confines of past practice. They must develop dynamic new approaches to modeling the effects of those tools on human decision making, and also the effects of human decision making on how those tools are used. To do so, however, designers must be given the time to design, not just to assemble the next game. They must also return to the fundamental principles of wargame design.

In addition, the designers of professional wargames can turn to the wargaming hobby for new ideas and new approaches. Although the details of hobby-game systems are seldom accurate enough to transplant whole to the professional world, it is more often the case that broad approaches developed in the hobby can contribute significantly to new professional directions. This is so because the fundamental principles of game design apply equally well in both arenas.

FUNDAMENTAL PRINCIPLES: A VIEW FROM THE HOBBY

In his book *The Complete Wargames Handbook*, James Dunnigan characterizes a wargame as a "non-linear communication" device. He also describes his two basic rules for constructing such a device. Expressed with typical, Dunniganesque irreverence, these rules are "keep it simple" and "plagiarize."[1]

Coming from Dunnigan, one of the hobby's great originals, the second rule seems so unlike his actual practice that it could easily be dismissed as mere flippancy. Yet Dunnigan is serious, and "plagiarize" is but a brash way of stating an important truth. In the long history of wargaming, or even the shorter but more intense history of hobby board wargaming, some fundamental techniques have proven themselves successful time and again.

For example, nearly every hobby wargame rates each unit for the number of hexes it may move in a single turn, and represents the effects of different types of terrain by imposing added costs in movement points for entering certain hexes or crossing certain hex sides. The vast majority of ground combat games are based on a comparison of attacker's strength to defender's strength, expressed as an "odds ratio." In naval games, even the most

188 elaborate systems are sometimes only variations on the basic idea of comparing offensive firepower to defensive "protection."

Even a revolutionary game system can make use of such proven techniques to simulate some aspects of reality. Not only does such judicious plagiarism take advantage of hard-won experience, but it may also make the player's task of assimilating a new system easier. An experienced wargamer who already understands the basics of some of a new game's "borrowed" subsystems can concentrate on adapting to the more revolutionary concepts reserved by the designer for the game's most important elements.

The designer who rejects this rule without good cause can, indeed, produce dramatically new and different game systems, but may pay a high price for doing so. There are numerous examples of hobby games in which many of the major subsystems used a unique, new approach. For example, Jack Radey's *Black Sea*Black Death* introduced brilliantly conceived new approaches to initiative, movement, combat, and the interplay of different types of forces and arms. Despite critical acclaim for its innovations, there is little evidence that the game is actually being played. It was, perhaps, too much of a good thing.

The key to understanding the importance of Dunnigan's second rule is appreciating the difficulties of achieving his first rule. Keeping things simple, especially when developing innovative approaches to game design problems, is one of the proverbial "easier-said-than-done's." As anyone who has ever tried to design a game can verify, it is far too easy for the game to take on a life of its own and explode beyond the confines of the designer's initial concept. There is a seductive tendency for a designer to become tangled up in micromodeling reality in new and alluring ways. When such subsystems are examined apart from the game as a whole, they often appear both reasonable and elegant, but when combined with all the other similar components that grew to maturity during the designer's quest for innovative ways of modeling the real world, the entire system becomes undecipherable, unwieldy, and totally unplayable.

As if to help himself follow his two precepts, Dunnigan seemed to practice his art at breakneck speeds. He churned out game after game, developing his technique in incremental stages rather than laboring for years over the last detail of the elusive "perfect game." To facilitate the assembly-line-like approach that dominated the early years of Simulations Publications, Incorpo- **189** rated, Dunnigan followed what he called the ten steps in the game design process. These steps, also described in *The Complete Wargames Handbook*, are:

1. Concept development
2. Research
3. Integration of ideas into a prototype
4. Fleshing out the prototype (adding the "chrome")
5. First draft of the rules
6. Game development
7. Blind-testing
8. Final rules edit
9. Production
10. Feedback

It is especially interesting to note that Dunnigan included the feedback step in his design process. The production of a game did not end the process of its design, at least not in the evolution of its designer's thought processes. Dunnigan believed that feedback from those who actually played the game was an important contributor to that evolution.

From his years of design experiences and the feedback and criticism of thousands of players, Dunnigan distilled two basic concepts that he believed must underlay the design of historical games. First, the game must accurately simulate, in some sense, the historical events it intended to portray. Second, the game's designer must carefully choose his focus and level of simulation detail and design the entire game to be commensurate with that choice. If a game dealing with the Nazi invasion of the Soviet Union failed to allow the re-creation of the spectacular initial German successes and their equally spectacular later reverses, it violated the first principle. If the game cast the players as the overall theater commanders and simultaneously required them to choose the assault tactics for individual corps or divisions, it strayed from the proper focus.

In these two basic concepts, Dunnigan succeeds in stating the two fundamental requirements of any wargame: realism, the accurate simulation of events in some sense, and playability, achieved by choosing that sense and also the focus required to make it meaningful. Achieving these goals demands a blend of controlled innovation and reliance on proven techniques. To

190 carry the process through, the designer must first develop the broad outlines of what he is trying to achieve, collect the facts necessary to inform and support his modeling, integrate his concepts and facts into a working model, and refine and document that model.

FUNDAMENTAL PRINCIPLES: THE PROFESSIONAL PERSPECTIVE

Unlike those in the wargaming hobby, professional wargamers work in a relatively closed society. One organization's games are not freely available for all to try, critique, and modify. Professional wargame designers may document their games (usually in classified publications), but they seldom describe the design process they employed to create them. The handful of open-source books that deal with professional wargaming (Brewer & Shubik, Hausrath, Wilson, Allen, and others) discuss the various types of games and the need for mathematical models to represent combat activity. For the most part, they fail to address the problem of actually turning a collection of such models into a game. It is this step that is the essence of the game designer's art.

In designing a professional wargame, the designer must ask and answer a set of questions that is somewhat different from those of greatest importance to the hobby designer. While the latter is concerned primarily with transmitting his interpretations of a situation to the game's players in an entertaining and educational format, the former must do more. The professional wargame must be designed to allow the players not only to learn, but to teach as well. The key questions for the designer of professional wargames are the following.

- What does the sponsor want to learn from the players?
- What information does the sponsor want to convey to the players? Specifically, in educational games, what does the sponsor want the players to learn?
- Who are the players that should be involved in the game, who are players that actually will be involved in the game, and what are their interests or concerns?
- How can the game best assure that the goals of the sponsor and the goals of the players will reinforce each other?
- In particular, what information must the game provide the

players, and what information must the play of the game **191** produce to meet the sponsor's goals?

- Finally, how can the game be structured to make the necessary information exchange possible, likely, and efficient?

Frequently, designers of professional games are drawn from the ranks of active duty or retired military personnel, computer programmers, or operations research analysts. Particularly for those in the latter two classes, game design presents some insidious pitfalls. Programmers and analysts have a tendency to address large and complex problems by breaking them down into smaller, more manageable pieces. The analysis of each of the subproblems is then integrated to address the overall issue. In the case of the usual types of formalized problems dealt with in computer programming or operations analysis, the links among the various subproblems can be specified and controlled by the analyst himself, and the unforeseen excluded, almost by definition. The unwitting game designer who concentrates on getting the pieces right and expects them to fit together to form a workable system almost as a by-product of the detailed modeling is more than likely headed for a rude shock the first time the game is played.

Models are not the game; they are not even the game system. They are only small components of what is a complex and dynamic communications process. To understand the fundamentals of game design, it is necessary to explore that communications process in greater detail.

COMMUNICATION AND THE ART OF WARGAME DESIGN

We have seen that whether it is a hobby game or a game for professionals, a wargame is primarily a communications device. In a hobby wargame, the communication is nearly always one-way, flowing from the designer to the player. In a professional wargame, on the other hand, communication is more complex, flowing from a sponsor through the designer to the players, and from the players, through the analysts, back to the sponsor. In either case, to understand the fundamental principles of game design, the prospective designer must have a grasp of the basics of communication.

Without getting too elaborate or philosophical, it seems clear that communication requires someone (the sender) who,

192   for some reason (the objective), wants to say something (the message) to someone else (the receiver). To do so, he requires some means (the medium) to carry the message, and some procedure (the process) to translate the message into a form compatible with the medium. If a response is required, the entire system must be able to work in both directions.

In a professional wargame, the sponsor operates through the game designer to achieve either research or educational objectives (or both) through the game. The participants of a professional wargame usually include not only the players, but a host of observers, analysts, and "expert witnesses," who are there to help the players or the control team and, incidentally, to learn what they can. The message flow includes the information provided to the players in the game's scenario and data base, and the decisions and feedback of the players. The tools used in the process of game play and analysis include mathematical models, control procedures, and the overall style of the game.

In a hobby wargame, the designer's goals are usually some combination of educating and entertaining his players (and the almost inevitable game reviewers). He, too, must provide his players with information and decision-making opportunities, but he receives only limited immediate feedback. As in professional games, the tools include the models, procedures, and style of game play. Table 2 compares and contrasts hobby and professional games as communications systems.

THE FUNDAMENTAL PRINCIPLES OF WARGAME DESIGN

Of Dunnigan's ten steps in the game design process, only the first five can be considered part of game design proper. The remainder can be considered elements of the development or analysis processes, which will be described in later chapters. Although not identical to Dunnigan's first five steps, the fundamental principles of wargame design are similar in many ways:

- Specify objectives
- Identify players, their game roles, and the decisions they will be expected to make
- Determine and collect the information they will need to make those decisions, and, for professional games, identify the information feedback required to achieve the game's objectives
- Devise the tools needed to make the process work

Table 2. Games as Communication Systems

System Element	Professional Games	Hobby Games
Sender	Sponsor, designer	Designer
Objectives	Education, research	Entertain, inform
Receivers	Players, observers	Players, reviewers
Messages	Information, decisions	Information, decisions
Medium	Models, procedures, style	Models, procedures, style
Process	Play, analysis	Play, review

● Document the results of the effort

Let us now look at each of these principles in greater depth.

Specify Objectives

Clearly, in both hobby and professional game design, specifying the objectives of the game is fundamental to the design process. In hobby games, the designer is usually relatively free to set his own goals. In professional wargaming, this initial step is one in which sponsors, designers, and game analysts must work together closely. They must not only identify the game's objectives, but also define how and in what ways the game will help meet those objectives. Often, the sponsor's initial goals will be unclear, or the utility of gaming for achieving those goals uncertain. The designer must play a major role in helping to sharpen the definition of goals, and the game analysts must help identify what gaming can and cannot contribute. Once the sponsor, designer, and analysts have agreed upon the definition of the problem, and decided how it may be usefully addressed through a wargame, the actual design work can begin.

The reasons for designing a hobby wargame can be almost as numerous as there are game designers. However, virtually all hobby wargames are intended as some form of intellectual entertainment, with a strong educational component. Most published hobby games are marketed among a small and relatively sophisticated audience having strong interests in military history and current military affairs. The subject of a particular game is usually chosen because the designer has some interest in a certain period

194 of history, a particular campaign, or a specific battle. Sometimes the designer may choose a subject specifically to showcase a new design concept (the so-called "system in search of a game"). In addition, many games are designed by hobbyists for their own use, or for the use of a small group of gaming acquaintances. Few such amateur games are ever published, although some of the more dedicated and ambitious gamers have succeeded in finding enough capital to produce their own game. Others may find an existing publisher willing to take on a free-lance design. Such "labors of love" result from the designer's abiding interest in a subject, or the designer's enjoyment of the design process itself.

In the professional world, the reasons for designing a wargame are usually expressed in terms of two broad categories of objectives, education and research. Educational objectives may be further characterized as focusing primarily on providing an active learning experience of their own, reinforcing lessons taught in a more traditional academic setting, or evaluating the extent to which students have assimilated such lessons. Research objectives may also be divided into three main classes; they may focus on developing or testing strategies and plans, identifying issues, or building a consensus among participants. Table 3 summarizes the various classes of professional game objectives.

Whether research or educational, professional or hobby, each wargame has its own unique set of objectives. These objectives are a blend of the various types and classes described above. As a result, although it is easy to state that a game's objectives play a critical role in its design, it is much more difficult to describe precisely how that role manifests itself in any general way. There is no recipe for translating a game's objectives into its mechanics. The principles described in the remainder of this chapter give some hints about the process, but ultimately the designer's talent dictates how and how well the translation from objectives to mechanics works.

Identifying Players and Decisions

Know Your Audience

As any writer, speaker, or performer is sure to understand, one of the first principles of communication is "know your audience." In a very real sense, the players of a wargame are its audience. Its

Table 3. Professional Wargame Objectives

Educational Games	Research Games
Teaching new lessons	Developing/testing strategy and plans
Reinforcing old lessons	Identifying issues
Evaluating students	Building consensus/understanding

objectives are the messages to be sent and received through the medium of the game, and the sponsor is the initiator of the communication. The game's players are the sponsor's correspondents in the process of learning and communication. Understanding who the players of a wargame may be or should be is thus a crucial step in the process of designing the game.

One frequent source of confusion in discussing games and game players is the fact that the term *player* actually has at least two popular meanings. Each individual decision maker in a game is called a player. On the other hand, the term is also used to refer to the numbers of independent sides represented in the game (for example, the usual "two-player" game of Red against Blue, or the "multi-player" games so frequent in the political-military arena). Finally, the term player may even be applied to the various game roles filled by participants (such as the "Blue-NCA player"). The designer must understand each of the meanings of the word and the distinctions between them. He must also distinguish between the roles and activities of actual players (decision makers) and those of the control staff and umpires.

The first step, then, is to know the people (or at least the characteristics of the people) who are likely to play the game. In a hobby game, the designer can often target a particular audience. He can design the game for those unfamiliar with or new to the hobby, or for the hard-core experienced wargamer (known in the hobby as a *grognard*, a French term meaning "grumbler" and a popular nickname for veterans of Napoleon's Grande Armée). Some hobby games are designed to appeal to a relatively wide audience; others are designed for special-interest groups who concentrate much of their gaming on particular periods or types of warfare.

Professional game designers seldom have the same luxury. It is often the case that a game's objectives can best be achieved if

196 particular individuals are involved in particular game roles. For example, a game dealing with the operations of a navy fleet may achieve its best results when the fleet commander "plays himself" in the game. Unfortunately, it is just as often the case that such individuals are not available to play. Thus, the professional game designer must temper his design by a realistic estimate of the background and experience (and rank) of those people who are likely to actually be assigned to play the game.

To achieve the full benefits of a wargame's ability to thrust a player into a simulation of a real-life operational role, the designer must identify those roles most important to achieving a game's objectives. He must make the roles given to the players consistent with the geographic and operational scope and scale of the game. The most successful game designs provide players with just such well-defined operational roles. The scope and scale of the game gives them geographical and operational command responsibilities commensurate with those roles, and the format and style of the game provides them with the types and quantity of information appropriate for carrying out their assigned functions. For example, in an actual situation a fleet commander may usually receive reports from and issue orders to his battle group commanders. In a navy wargame of such a situation, the player representing that fleet commander should do the same.

The simplest example of the kinds of problems that may result when player roles are not carefully structured to reflect actual operational responsibilities can be found in an all-too-large number of two-player hobby board games in which players are required to represent the supreme command of an entire nation at war, but are also required to maneuver individual divisions or ships in each theater of conflict. Such games force the players to spend so much of their time worrying about detailed tactical options that they must give short shrift to the strategic considerations on which their real-life counterparts would spend most of their time. Of course, the players could attempt to give adequate attention to all the various layers of command functions the game's design imposes on them. Doing so, however, almost inevitably results in the game's play dragging on for inordinately long periods of time. Typically, the hobbyist's solution to games that suffer from this affliction is to bring in more players to divide the load (as was done in the *War in the East* game described earlier) or to let the game gather dust on the shelf (the more frequent, and unfortunate, solution).

The decision-making level of the players is inextricably linked to the geopolitical scope of the game and to its scale of aggregation (both geographic and organizational). All of these elements must, in turn, be consistent with the game's objectives and with the number of player sides and the number of people actually involved in representing each side.

Because most professional wargames are so detailed, complex, and specialized, it is rare that a single individual can play a side in the game. Typically, teams of several officers staff what are usually referred to as game "cells." Such cells could represent the watch team in a combat information center, the battle staff of a fleet commander, or the National Security Council advising the president. The actual number of people involved in playing a cell depends on the information processing and communications requirements the game levies on the cells, and on the wishes and resources of game sponsors, players, and facilities.

The key point for the designer of professional games is to assign player cells to command positions of most interest and greatest importance to achieving the game's objectives. Command levels above and below these can be assumed by the game control staff. For example, a game focusing on theater-level operations may demand player cells at fleet and battle-group levels, with the control staff assuming the roles of the National Command Authority and of individual ship commanding officers. The difficulties of finding a single opponent, much less someone willing to serve as an umpire, handicap hobby games in this respect, and are often the source of the difficulties that plague the definitions of player roles in such games.

Keep Players Out of the Weeds

Another problem is the natural, indeed the inescapable, tendency for a game's designer to include more detail than is really appropriate for the command levels he is trying to represent in the game. What can make this problem worse, especially in the hobby arena, is the fact that there is often an even stronger tendency on the part of the players to desire greater and greater levels of microscopic details. It is a well-known phenomenon among hobby game designers that many, if not most, players want to ex-

198 hibit their tactical skills during the play of the game, and are not content with demonstrating their strategic acumen. For such players, it is not enough to order two battle groups to strike an enemy-held island—they often want the pleasure of choosing the individual squadrons to make up each attacking wave and the strike routes those squadrons will take.

A similar phenomenon can often be observed in the play of navy wargames. The difficulties of getting senior officers such as fleet commanders to spend the time required to play a wargame often result in using junior officers to play such roles. Usually, these officers are unfamiliar with their game roles, and sometimes may even be uncomfortable with those roles as a result of that unfamiliarity. Unless the game's design and control group work to overcome this problem, players may naturally begin to exhibit more and more interest in aspects of the game with which they are more familiar and comfortable.

This phenomenon of "getting lost in the weeds" is usually not only inappropriate to the players' game roles, but often disrupts the flow of the game. When a player who is supposed to represent a fleet commander spends inordinate amounts of time discussing the ingress routes for a flight of tactical strike aircraft or arguing about which evasive tactics such aircraft should employ when bounced by enemy interceptors, it not only slows down the pace of the game, but also prevents him from devoting the careful thought to higher-level strategic issues that his role should require.

Once outside the game environment, designers and control personnel frequently decry these tendencies. During actual play, however, they too often feel compelled to bow to the player's wishes. Sometimes such lower-level discussions and debates are useful and even crucial to achieving a game's objectives. More often, however, success of a game will depend strongly on avoiding too many such diversions. The first step in solving the problem is identifying its true source.

Typically, players begin operating outside the bounds of their game roles as a result of a failure of context and a subsequent decrease in player enjoyment of the gaming process. Unless a particular command role is, in reality, uninteresting or unchallenging, reproducing it will provide an informative and engrossing game experience. If the re-creation of roles is poor, however, it can leave a player with too little to think about and do, and so force him to look for "fun" in other directions. Unless a game can avoid this problem completely, it must provide alternative means of informing and diverting players to prevent them from mucking about in the details of play. In professional games, one frequently used approach is to schedule a series of seminars to provide the players useful information during the periods when they might otherwise find too little to do.

Give Players Synthetic Experience

There is another side to the coin of grasping and holding a player's interest in performing his game role. It is simply not much fun and seldom very informative for a player to be required to do things he simply does not know how to do. If a player is unfamiliar with his game role, the designer must design the game in a way that helps the player carry out his role competently. The key to doing this lies in providing players a level of information and a structure for game play that gives them, in a sense, synthetic experience in their game roles.

The structure of the game provides the player with a framework for understanding what decisions he must make, what factors he should consider in making those decisions, and what form the decisions should take. The information provided to the player should be organized in a way that gives him a sense of the possible effects of the important factors, along with enough extraneous details to make the task of sorting out precisely what is important sufficiently difficult to be realistically challenging and educational.

During its history, the board-gaming hobby has developed several approaches to providing players with enough synthetic experience to allow them to feel comfortable in their roles. The early games of the Avalon Hill Game Company (the so-called Avalon Hill Classics) shared very similar systems for movement and combat. Indeed, players of these early games such as *D-Day*, *Afrika Korps*, and *Waterloo*, easily memorized the Combat Results Table that dictated the outcomes of battles and was common to all the games.

As a result of their virtually identical basic systems, players of the Avalon Hill Classics found that certain fundamental tactics were essentially transferable among the games—that three-to-

200 one odds would guarantee the destruction or displacement of the enemy and the "five-to-one, surrounded" attack would automatically eliminate him. Similarly, the physical characteristics of the hexagon grid dictated how to achieve the required positions and concentrations of force. Once these fundamental tactics were learned, players could focus their attention on the strategic and operational contexts of the individual games. These latter could be discerned principally by studying the victory conditions and the differences in relative force capabilities, both between the opposing sides, and among the individual units. Sometimes it even helped to know a bit of the history of whatever operations the game represented.

Layered over this relatively straightforward underlying structure was the "chrome," the little details added to try to represent some specific, critical elements of the particular campaign or battle. For example, *D-Day* added rules for Allied amphibious operations and airpower, and German fortifications; *Afrika Korps* introduced detailed (for the time) supply requirements and units representing supply trains and dumps. These complications were attempts to introduce greater realism and provide the players with new challenges from game to game, but they avoided overtaxing the player's capacity to deal with too many new ideas at once.

The early Avalon Hill Game Company products remain prominent examples of Dunnigan's two principles of "keep it simple" and "plagiarize." They were extremely successful in their day, and remain popular even now, despite, or perhaps because of, their relatively simplistic treatments. Those early games created a hobby virtually from thin air.

Even in today's world of increasing complexity in its games and greater experience and sophistication in its practitioners, the wargaming hobby continues to find multiple games built around the same basic system an attractive commodity. This fact is reflected, for example, in the popularity of Game Designer's Workshop's *Third World War* and *Assault* series, Victory Games's *Sixth Fleet* series, and the Avalon Hill Game Company's *Squad Leader* and *Advanced Squad Leader* families.

Perhaps surprisingly, the company that Dunnigan built, Simulations Publications, Incorporated, took a different tack, one that seemed to apply Dunnigan's principles to a lesser extent than exhibited by the earlier Classics. Simulations Publications stressed the production of large numbers of very different games about diverse topics ranging from tactical combat on the Eastern Front of World War II (*Tactical Game 3*) to strategic operations in the ancient world (*The Fall of Rome*). The fundamental systems of such games, not surprisingly, tended to differ dramatically from one another.

To simplify the learning process, Simulations Publications tended to focus more on similarity of form rather than substance. Under the influence of art director Redmond Simonsen, Simulations Publication's games adopted a functional graphical style designed to simplify the players' information-processing problem. In many games, especially those published in *Strategy and Tactics* magazine, players could find a discussion of the historical background of the campaign or battle represented by the game to give them an understanding of its real-life context. There were often designer's and player's notes provided to help the player understand the rationale behind the game system and to provide him with some helpful hints on playing the game.

While successful for a time at increasing the numbers of games available to hobbyists and encouraging design innovations, the Simulations Publications approach was less successful at bringing new players into the hobby's ranks or at allowing experienced players to develop greater expertise. Instead, the number and diversity of increasingly complex games led to the formation of small, often isolated, pockets of specialists who could play one or a handful of games very competently, but were totally out of their element when trying to play something different. Experience became less and less transferable.

The end result of this process was that most games produced by Simulations Publications, Incorporated were not played by most people, even by those who bought the games through subscriptions to the *Strategy & Tactics* game-in-a-magazine. The new approach had responded to player demands for more games, on more topics, in more detail, but it had failed to help the players assimilate what was needed to play the games well. Wargames became even more elitist, and more work than fun. The almost overwhelming levels of complexity of some of Simulations Publications's later games exacerbated the requirements that players adopt a multitude of different command levels and roles during game play.

202   The demise of Simulations Publications, Incorporated in the late 1970s was followed by an increasing realization on the part of game designers that realistic games could be designed so that players could and would play them, but only by stressing their role-playing aspects. Many computer wargame designs, prominently those of Australia's Strategic Studies Group's *Carriers at War* and *Europe Ablaze*, and Simulations Canada's *Fifth Eskadra* and *Seventh Fleet*, have stressed this "viewpoint oriented" approach to design.

      Unfortunately, computer-game designers, perhaps even more so than their board-game counterparts who have a long-established body of experience, often fail to provide players with the synthetic experience needed to play the games competently. Often this failure results from the designer's overly conscientious attempts to conceal from the player information that, in real life, participants would not know, or at least not know precisely. Computer-game designers rightly resist providing the players with precise and detailed knowledge of the parameters and inner workings of the game system and its models of reality. The design failure lies in opting to provide players no information or "helpful hints" at all. As a result, frustration levels can be high as players attempt to learn what does and does not work in the game, suffering one embarrassing defeat after another at the hands of the machine. While it is difficult to argue with a philosophy that says players should not know more than their real-life counterparts could, designers must also remember that most players have not reached their command positions only after the years of operational and exercise experience enjoyed by their real-life counterparts. The designer who fails to provide his players with the synthetic experience needed to play the game has not designed his game properly.

      From the player's perspective, the play of the game must be challenging, involving, and educational, but not impossible and frustrating. Because the players are the critical elements of a game, the game's designer must build his game to help the players suspend their disbelief, much as a novelist would. But the game designer must go even further, getting his players to contribute actively to building and participating in the fiction.

      In order to accomplish this, the designer must, of necessity, make the players want to play the game, or at least want to play it well once they are involved. To accomplish this, in turn, the designer must work to help the players play competently, if not well, as quickly as possible. He cannot afford to allow the players to feel that they simply cannot cope with the demands the game places on them, or they will simply stop playing. Thus, the designer must understand his players, devise a means to help them suspend their disbelief, provide them with the synthetic experience they need to play reasonably well, and also give them the ability to assess their performance during game play. He accomplishes these goals by providing the players with the information they need, the mechanics for turning that information into decisions, and the outcomes of those decisions.

Define Information Requirements

Good wargame design should enable game designers, sponsors, players, analysts, and other participants to communicate with each other, often in an iterative process. The designer begins this process by placing the players in the midst of a situation that requires them to make decisions. This setting is commonly referred to as the game's scenario. In addition to the setting, players must also receive information about the people and objects their decisions will affect. They must have data allowing them to estimate friendly and enemy capabilities, levels of supply, and other important elements of forces and physical conditions. They must also be able to form some judgment about what might be the possible outcomes of various decisions they may choose to make. These latter types of information comprise the game's data base. Finally, there is other information, in the form of updates on the status and operations of friendly and enemy forces, that arises during the play of the game.

Scenarios

The term *scenario* has its origins in the theatrical world, where it usually refers to an outline or synopsis of the plot of a play, novel, or other work. Wargame scenarios set the scene for player decisions and provide for specific updates in the situation during play in order to alter or influence the developing situation and to elicit player responses to specific items of interest. By defining the set-

204 ting and scope of player decisions, scenarios can direct the course of a game into either very narrow or fairly broad channels, depending on the game's goals.

For example, a tactical, educational game dealing with the employment of surface-to-surface missiles may use a simple scenario in which a single friendly ship is tasked to engage a single enemy ship known to be operating in a well-defined region. In such a scenario, geopolitics, war aims, and other deep strategic considerations may be superfluous. A research game exploring warfighting and strategy issues for a global conflict will require a scenario focusing on just such high-level factors. Player decisions in the first case will be limited to those required to maneuver against, target, and destroy the opponent with surface-to-surface missiles. In the latter game, players may have far greater latitude in choosing the theaters of military action and the forces committed to those theaters.

Because the guidelines, bounds, and subtle influences of scenarios can become straitjackets to players' decisions, the game designer must be sure that the scenario allows the players enough decision-making flexibility to let the game meet its objectives. Because a game's objectives should focus on exploring the factors and reasoning that affect specific types of decisions, scenarios should be designed to minimize artificial restrictions and allow players as much freedom of choice as possible in making those decisions.

For example, one goal of a game may be to study the factors affecting the relative allocation of navy forces to the defense of sea lines of communication and to forward offensive operations. In this case, the scenario must at least define a potential threat to the sea lines of communication and a potential benefit to offensive operations so that choices between the alternatives may be made for strategic or operational reasons, not merely by default. Similarly, if a game seeks to explore operations in a global war with the Soviet Union, scenarios that assume away the existence of tactical nuclear weapons may introduce severe biases in player decisions. With no possibility of a nuclear threat, players may operate their forces in ways that are highly unlikely if the threat of escalation is a real one.

In hobby games, scenarios typically define the area of operations, the forces involved on both sides, and the victory conditions or objectives. In historical games, particularly those dealing with situations that saw one of the combatants wield a decisive and overwhelming advantage, designers often include hypothetical or "what-if" scenarios to provide players with a more balanced, competitive game. Often, these hypothetical scenarios assume that some critical decision taken before the time represented in the game had gone a different way. The designer then projects some estimate of the effects of this change of decision, and designs his scenario to take these effects into account.

One striking example of this idea is seen in *France, 1940.* In this game of the German blitzkrieg against France and the Low Countries, the historical scenario proved as much of a cakewalk as the actual events. To make for a more interesting game, designer Dunnigan included a wealth of what-if scenarios, ranging from a French decision to forgo the Maginot Line in favor of creating more armored divisions, to a commitment of greater British air power to the campaign.

Designers of professional wargames have fewer opportunities to allow the same group of players to replay a game under different scenario assumptions. Nevertheless, research into the roles, desirable characteristics, and evaluation of scenarios for professional wargames (and other types of research, such as systems analysis) is ongoing.[2] For game designers, the most important elements of this research deal with identifying some of the basic components of a scenario and defining some fundamental principles of scenario design.

Simply stated, a scenario should include all essential information about the game's setting and subsequent planned modifications to it, and should contain no superfluous information. For example, a game dealing with submarine tactics in the Arctic will seldom require a scenario that defines the balance of land-based tactical air power in the Indian Ocean. The trick for the designer is to determine what is essential and what superfluous.

The scenario is the common starting point from which sponsors, players, analysts, and other game participants address the goals of the game. As such, the scenario must provide a description of the context or background from which it arises, including the general physical and geopolitical situation. It should also include the attitudes, goals, and intentions of the actors involved, whether friendly, enemy, or third-party. Of course, such

206 information should be limited in quantity and quality to reflect real-world uncertainty and inaccuracy.

Hobby games, especially those dealing with the interaction of politics, diplomacy, and military force, often include some provisions for the actions of the players affecting the attitudes and activities of non-player nations or forces. Designer Richard Berg is perhaps best known for his use of this device, typically incorporated in a table of "Random Events" that may occur to plague (literally) the unsuspecting player. One of his early games, *Conquistador*, dealt with the European exploration of the New World and included events such as:

- Costly European War Drains Treasury
- Graft and Corruption Rife
- Guile and Treachery Obtain "Rutter" (a Rutter was a set of sailing instructions for traveling to a particular location; in the game, players need a Rutter to sail their ships around Cape Horn)
- Natives Resentful of European Intrusion.

Other games, such as *Third Reich*, allow players to choose some possible modifications to the historical situation, and the interactions of their choices can have specific, and dramatic, effects on the game conditions.

In *Third Reich*, ten numbered chits were used, from which the British player and the German player each chose one, which could be played at some time during the game. The meaning of the chits differed for each side. For the Germans, chit number 7 read, in part "Turkey enters war on Axis side as German Minor Ally provided Germany has invaded Russia and currently has an advantage in combat factors on the Eastern Front." For the British, the same chit read "Allied Strategic Bombing concentrates on German crucial industry," a contingency that doubled the effectiveness of Allied strategic bombers. More interestingly, because there was only one chit of each number, "if the British player selects option No. 7, he knows and can relay to his Russian ally that there is no danger of a Turkish invasion because the German will be unable to pick his own option No. 7. This can be chalked up to successful espionage and intelligence efforts."[3]

In addition to background information, scenarios should also include guidance to each player or cell about the specific objectives or missions that those players may be called upon to pursue. Command relationships among players and cells and between players and control should be clearly spelled out, as should the assignment of forces and support. When updates to any or all of these elements of a scenario are planned, such updates themselves should be considered part of the scenario, though obviously they should not be provided to all the players beforehand. Similarly, if the control team must respond to player actions or requests (such as requests for nuclear release) in specific ways for the game objectives to be met, such instructions are considered part of the scenario.

Because most hobby board games are designed to be played by a single person on each side, the modeling of command relationships is seldom needed. Some games, however, have included command subsystems to restrict the activities of units in the game in order to more accurately reflect the limitations of the real world commander, despite the virtual omnipotence of the player. Such systems (*Frederick the Great, Terrible Swift Sword,* and *Assault,* for example) have been applied to all levels of warfare. Miniatures gamers have also developed elaborate systems of command and communications for their multiplayer affairs.

Hobby games have a fundamental difficulty with updating scenario conditions during play, again because of the all-knowing position most players enjoy. The major exception to this problem arises with computer games. In games like *Seventh Fleet,* the player never knows when his superiors will begin the war or even escalate it to nuclear levels. Without such use of computers or human umpires, hobby games are limited in the scenario flexibility they can achieve. The control staffs available to most professional games alleviate much of this problem.

Principles of Scenario Design

The primary components of a wargame scenario are listed below. Designing a scenario consists largely of tailoring the components listed to create an environment in which the objectives of the game can be met. Good scenario-design practice involves four fundamental principles: understanding the problem, building from the bottom up, documenting choices, and communicating results. The first and most basic principle—understanding the problem—is perhaps the one most frequently violated, possibly because it is so obvious. As with everything else in a game's de-

208 sign, the scenario begins with the game's objectives. But just understanding the objectives is not enough; the scenario designer must also understand how those objectives are to be met in the game. He must identify the kinds of player activities and decision-making opportunities that are required to meet the game's objectives, and then he must ensure that those activities and opportunities can arise.

Components of Scenarios

- Background
    Situation
    Attitudes
    Intentions
    Goals
    Physical conditions
- Objectives or missions
    All players and cells
- Command relationships
    Among players and cells
    Between players and control
- Resources
    Force structure
    Available support
- Updates during play, and control team instructions

For example, in the surface-to-surface missile game described earlier, possible game objectives of teaching proper radar-targeting procedures require that the target can be detected. The scenario writer must choose a physical and tactical environment to ensure this possibility. Placing the target close to a shoreline so that it is difficult to distinguish its radar reflection from the land clutter may teach valuable lessons about real-world tactical possibilities but will make it difficult to achieve the radar-targeting goals.

Once the game's objectives and the means for achieving them are thoroughly understood, the scenario writer may begin to structure the flow of game play to allow the means and ends to come together without forcing the players to follow a single, rigid path. A critical node in the flow of a game, at which a player's decisions will lead inevitably down one or another of several major alternative paths (to order a battle force into a fjord or to hold it in the open ocean; to launch a strike immediately or await the enemy's attack first), may be called a decision point. Such decision points must be defined in enough detail to allow players to identify a realistic range of alternatives while restricting that range to a reasonably limited type or number so that the game's control personnel can be adequately prepared to evaluate any player decision.

The key to this process is a bottom-up hierarchical approach. The designer begins by identifying the specific sorts of decision points required to meet game objectives. He then must step back in time to determine the possible sequences of events that could lead to such decision points. In this process, the designer seeks to identify critical events, decisions, or actions that are required to reach the particular decision point and that are beyond the control of the players. Such events are then incorporated into the scenario. In many cases, these considerations shape both the scenario updates during play and the control-team instructions.

As an example, consider a professional wargame designed to explore global warfighting issues. A critical issue in any global war is the decision to employ nuclear weapons in any form. It may be the case that allowing such weapons to be used at all would be contrary to achieving many game objectives or would make the combat-evaluation tasks of the control personnel too difficult or speculative. One solution to this problem, and unfortunately the one most frequently employed, is for the scenario to explicitly tell the players that nuclear weapons will not be used. A better alternative might be to acknowledge their existence and possible employment in the scenario background information provided to the players, but to instruct the control team to refuse nuclear release should players request it. The overall effect, that nuclear weapons are not used, is the same. However, the latter approach will allow both the investigation of conditions under which nuclear release might be considered and the exploration of effects such possibilities might have on planning and operations during the period of conventional warfare.

The bottom-up approach allows the scenario designer to build in and constantly monitor the completeness, coherence, and credibility of the scenario. A complete scenario provides all those involved in the game and its analysis with the information they require to carry out their roles. If a theater commander is not given a mission, the scenario is incomplete. If an analyst is

210   not told why the war begins, the scenario is incomplete. In addition, the completeness of the scenario allows participants and future students of the game to "appreciate the objectives and scope of the . . . [game] as well as the subtle issues it poses,"[4] and to separate those issues built into the game by the scenario from those issues arising during the play of the game itself.

Coherence of scenarios means that the scenario assumptions must be logically consistent, but it implies more than mere consistency. A coherent scenario ties together all the elements of the game, from its objectives, to its mechanics, to its analysis. If an important objective of the game is the exploration of under-ice antisubmarine warfare operations, a coherent scenario will assure not only that such operations will take place, but also that they will take place in ways that can be handled by the game's mechanics and that can be recorded and interpreted by the game's analysts.

Finally, a scenario must be credible in the sense that game participants and later audiences for its results are willing to suspend their inherent disbelief in hypothetical situations. The scenario represents a view of a possible reality. That reality need not be the most likely one, but should generally be a possible one. As Carl Builder suggests, the starting point should be current reality, and the development of the scenario's fiction should proceed logically from that reality according to the documented decisions and assumptions of the scenario designer.[5] Some elements of the scenario world may be perceived as extremely unlikely, if not impossible (as for example, the games played at the Naval War College between World War I and World War II that pitted the U.S. and British fleets against one another). However, if those elements most applicable to the objectives of the game are perceived as important and in need of exploration, the suspension of disbelief can still be achieved. (The need to study tactical alternatives against a superior fleet using existing systems allowed players to overcome their inherent skepticism about a U.S./U.K. war at sea.)

Throughout the scenario design process, it is important for the designer of a scenario for a professional wargame to document his decisions. He should record the reasons behind his choices of assumptions, the factors included or excluded from the scenario, the use of particular sources of information, and any other decisions he makes. Thorough documentation allows the designer to be sure of just what went into the game, especially if there are likely to be questions about what comes out of it. And thorough documentation provides a solid basis for the final and (in many ways) most crucial elements of the scenario process.

To be of any use, a game scenario must be communicated to the people who will use it. There are basically five classes of users: game sponsors, game-control personnel, game players, game analysts, and future audiences for game reports or other summaries. Sponsors need to be sure that the scenario will facilitate meeting the game's objectives. Control personnel need to understand the context of the game and the prerogatives and limitations under which they must operate. Players need the same information as control personnel, but it should be constrained to reflect their less-than-perfect knowledge of their game world. Analysts need to know not only the full story of what and why, the "ground truth," but also the story as told to players and control, so that they may interpret the effects of information constraints. Finally, the future consumers of the game's issues must know not only the context of the game, but also how to distinguish scenario input from game-play output. Communicating effectively to such diverse groups requires not only literary and graphic skills, but first and foremost the complete information provided by thorough documentation, subsets of which can be tailored for particular groups.

The key elements of the scenario design process are listed below.

Elements of Scenario Design

- Understanding the problem
  - Game objectives
  - How scenario affects achievement
- Building from the bottom up
  - Define decisions points
  - Hierarchy of information and assumption
  - Completeness, coherence, credibility
- Documenting choices
  - Reasons for assumptions and decisions
  - Sources of information
- Communicating results
  - Sponsors, control, players, analysts, audience

Data Base

As described above, a wargame scenario contains largely qualitative information about the state of the world in which the game takes place. The data base of the game contains the quantitative information about capabilities of forces, levels of logistics, and relative likelihoods of the occurrence and outcome of interactions between forces. The data base links the scenario and the mechanics of the game. It must provide all the inputs required to allow the game's models to reproduce the qualitative scenario conditions and to generate outcomes of interactions.

But a good data base is more than raw, unprocessed inputs. Many current professional games suffer from the fact that players are provided a mass of raw data that contains much superfluous information. In turn, important data are difficult to find or require detailed calculations that are impractical for the player to carry out during the game. Players representing a battle force or fleet commander seldom need to know the cruise speed of a Harpoon surface-to-surface missile. On the other hand, they often want to know the chances of a successful Harpoon strike against a group of enemy ships.

The data base provided to the players should be tailored to their game role, to the types of decisions they should be making, and to the types of information they need to make those decisions. More detailed data are required by control teams and umpires and can be made available to the players if the situation warrants it. Furthermore, actual commanders are likely to have access to quantitative analyses of raw data that would give them estimates of capabilities or chances of success in many situations. The shortage of similar capabilities in a game makes the player's job much harder. The use of actual operations analysts to play their real-life roles in professional wargames is one solution. Another is the use of extensive preprocessing of raw data. Employment of either of these options makes for a more realistic and more efficient decision-making process. Preprocessing data can also greatly help in the suspension of disbelief. If players are aware of the range of possible outcomes of their decisions and have some idea of relative likelihoods, they will be more willing to accept an unlikely result as the "fortunes of war" rather than a dastardly plot by the control team.

Most hobby games are, by their very nature, almost immune to this problem because the players themselves must resolve all interactions and so must have access to the pertinent data. Computer games can relieve the players of this responsibility, but also deprive them of much of the data they need to make informed game decisions. As described earlier, this approach also deprives players of their needed synthetic experience and can lead to many frustrating experiences. In a very real sense, board games provide their players with the pertinent preprocessed data, and computer games typically do not.

A simple preprocessing technique that can be applied in many areas of both hobby and professional gaming involves developing graphs for various types of interactions. These graphs can display the sensitivity of outcomes to uncertainties in critical factors. Players can be provided with such graphs and can make decisions based on their perception of the values of the key parameters. This perception can be based on their own experience, on analysis, and on information provided in the data base, or on a combination of all of those. The actual interaction is then resolved using the same graphs but with inputs selected by the umpire (or computer) using pre-chosen values or pre-defined procedures.

For example, figure 7 shows a generic graph depicting aircraft attrition against different levels of air-defense capability. For each level of capability, a range of uncertainty is shown. The player may be planning to strike a target that his intelligence personnel estimate to have a low defensive capability. He can then determine that his losses are likely to lie between X and Y. If the strike is made and losses of level Z occur, instead of complaining about an unrealistic model, he may be led to believe that his intelligence people underestimated enemy defenses.

Of course, preprocessing data in this manner requires a great deal of preparation prior to game play. In detailed tactical-level games in which there are large numbers of parameters to vary over wide intervals, extensive preprocessing may be impractical. However, the computing resources readily available to most professional wargaming facilities are powerful enough that graphic displays for at least some aspects of game play can be developed fairly easily and quickly. Such an approach reduces the actual time spent playing the game by limiting the real-time use of

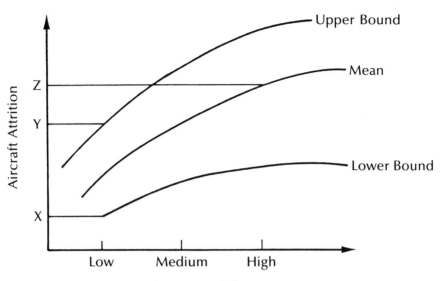

Air Defense Capability

Figure 7. An example of preprocessed data for wargames.

complex models, relegating those to the preprocessing phase of game design. Actual game play would then only require random number draws to access specific results, or special model runs for cases well beyond the bounds of the preprocessed data. Similarly, hobby computer games might be adapted to display some of their resolution routines in a similar form, allowing players some reasonable amount of information about what they might expect.

Devise the Tools

If the scenario sets the stage for the game, and the data base provides the information required for players to make decisions, the game's mechanics allow those decisions to be implemented and inform the players about their effects. Wargame mechanics may be considered as two interrelated systems: models and procedures. The models translate data and decisions into game events. Procedures define in game terms what players can and cannot do and why, sequence the game events to allow for accurate recreation of cause and effect, and manage the flow of information to and from players and control.

Models

Wargames use models as representations of all the aspects of reality the game may be required to simulate. These models may take various forms. Many nineteenth-century wargames relied largely on the judgment of experienced professional soldiers and sailors. Modern professional games often employ complex mathematical expressions programmed into high-speed computers. Modern hobby games use a combination of unit ratings and tabulated battle outcomes (combat results tables) chosen by comparing opposing ratings.

A professional wargame's models and the results they produce are best considered as inputs to the game, devices to move game play along rather than measures to evaluate player success or failure. Specific model needs vary from game to game. Some of the broad categories of models most likely to be required in wargames are the following:

- Physical environment
- Kinematics
- Intelligence collection and dissemination
- Command, control, and communications
- Sensors
- Weapons
- Logistics

No matter what the form or subject of wargame models, good ones share the following key characteristics:

- They accurately reflect factors most prominent for player decision levels
- They are flexible enough to deal with unusual decisions
- They are adaptable to changes in the data base
- They are stochastic to the extent reality is stochastic
- They are documented to allow others to understand assumptions and algorithms

Chief among these characteristics is the need to reflect accurately the influence of those factors most prominent in the decision processes of the game's player roles. Other factors, many of which may be important to the outcome of an actual event, are aggregated rather than assumed away. For example, a battle-force commander may determine the size and composition of an air

216 strike. Attacker and defender tactics during the strike, while critical to the outcome in the real world, are beyond the direct influence of the commander and so may be aggregated in the wargame's model of strike resolution.

There are two basic approaches to using models in wargames. The first approach employs detailed models to preprocess data before the game is played (as described in the previous section). Complex calculations or judgments are used to prepare tables and graphs of outcomes as simple functions of player decisions. These tables and graphs become the models actually used during the game play. This approach was popular in professional gaming before the advent of high-speed computers. For example, a wargame employed during the late nineteenth or early twentieth centuries might use data from fleet gunnery ranges, ballistics calculations, and measures of armor penetrability to devise tables showing the probability that a round fired from a battleship's main battery would disable the turret of an opposing ship at various ranges. This approach has the advantage of transparency (players can easily see what the relative odds of various outcomes are), ease of use, and direct applicability to player decision making. It has the disadvantage of being limited to pre-calculated cases, requiring umpires to interpolate or extrapolate from the given cases when circumstances in the game differ from the assumed conditions.

Modern computing capabilities allow professional wargame designers the option of using a second modeling approach. This approach uses the complex models to calculate results during actual game play. If a battle group is attacked by missile-carrying aircraft, the precise composition of the opposing forces and their sensor and weapon capabilities can be entered into a computer model. The model can account for battle-group formation, attack spacing, and many other parameters. It can calculate losses to aircraft and different types of damage to each target ship. The use of such detailed models has the advantage of appearing to better replicate the precise conditions of game play in the assessment of outcomes. Nevertheless, the price to be paid is that the players can no longer see the relative odds generated by the models and so find it difficult to incorporate expectations about outcomes into their decisions. This is the problem seen so often in the case of hobby computer games.

Another disadvantage of the use of detailed, complex models during game play is often a severe slowing down of play. This phenomenon can result from the need to provide detailed models with detailed inputs, which in turn require more detailed planning on the part of the players. Inevitably, such micro-modeling of reality results in more time spent on processing interactions.

Thus, when surveying the models available for use or defining those that need to be devised, the wargame designer must balance the advantages and disadvantages of the two modeling approaches. Part of the art of game design lies in choosing the best mix of approaches and models to maximize advantages while playing down the problems. Often, however, the design solution must go beyond the models themselves to the procedures through which those models are applied to the play of the game.

Procedures

Scenarios, data bases, and models are all necessary components of a wargame, but it is the procedures for their orchestrated use that set the game in motion and keep it on track. These procedures are specified by what are sometimes known as the game's rules, and in professional wargames are usually monitored by a team of umpires, referees, or controllers. These umpires and the procedures they employ have three main functions. These functions, derived from Frank McHugh's seminal work, are listed below.[6]

Functions of Procedures and Umpires

- Monitor player actions
  Translate player actions into game terms
  Enforce the rules of the game
  Prevent physically unrealistic actions or sequences of events
- Assess interactions
  Use models, data, and rules
  Use judgment when required
- Inform players about action outcomes
  Employ realistic limitations
  Introduce "fog of war"
  Umpires and procedures translate player decisions into terms that can be understood by the game's models. If a battle-group commander decides to launch an air strike, there must be a pro-

218 cedure for determining which aircraft are available for the strike and how long they will take to reach their target. The umpires must enforce the dictates of the game's rules and models. If the rules require a ship to lose its weapons capability after a missile hit, the umpires must ensure that this requirement takes effect.

Umpires and procedures must also prevent physically unrealistic actions or sequences of events. Ships may not sail over land, nor may a column of tanks cross a river before a bridge is built to carry them.

In carrying out these functions, however, umpires must avoid forcing the players to play the game the way the umpire or sponsor would have played it. Procedures must establish wide yet realistic bounds within which the players must be free to try their own ideas. This is especially true for most hobby games, which must be played without the benefit of an umpire, forcing the designer to be far more imaginative in estimating possible player actions and far more rigorous in defining procedures to resolve them.

Assessing the outcomes of game events is the second major task of procedures and umpires. Player decisions about the movement and use of forces, sensors, and weapons must be evaluated for the possible interactions they might cause. Typically, the results of such interactions are assessed using the game's models and the judgment of the umpires. Procedures for determining model inputs and interpreting model results are often required when special-purpose models are not developed for a specific game.

As an example of these effects, consider the basic system of battle-damage assessment used at the Naval War College for many of their major games in the early 1980s. The standard format for battle-damage assessment revolved around a small team (generally no more than five people) of battle-damage assessors, who used a combination of microcomputer warfare models and look-up tables to produce outcomes for all interactions in the game.

The method used by this battle-damage assessment (BDA) team is relatively easy to describe in superficial terms, but is more difficult to understand. Game controllers would initiate an assessment by bringing the BDA team an interaction sheet, which described the forces, dispositions, weapons, tactics, and any special conditions that applied to the engagement. The BDA team then used their models, data, and professional experience and expertise to evaluate the engagement and produce an outcome.

Sometimes the precise inputs to a model were skewed from those given on the interaction sheet to account for the peculiarities of a particular model, or for certain important operational or tactical factors not explicitly incorporated into the model's mathematics. Sometimes the outputs of a model were also skewed to account for the same sorts of factors. The final result of the assessment was then presented to the game director for approval. The director was responsible for ensuring that assessments were reasonable and also consistent with the sponsor's objectives for the game.

This approach to assessment mechanics differs from the rule-based systems typical of hobby board games. Although the assessors use rules, models, and charts as aids, they rely on their own knowledge, beliefs, and expertise to determine the outcomes of the engagements, and thus to control and direct the experience of the game's players. This "free-Kriegsspiel" style of assessment differs from the "rigid Kriegsspiel" philosophy, which controls the experience of the players through a rigid set of objectively applied rules and procedures that restrict player options and, within some small range of random numbers, strictly determines results.

The War College approach has several advantages over the apparently more objective and rigorous techniques seen more frequently in hobby games. The assessors can respond appropriately to nearly any action the players might take, even to unanticipated ones. This flexibility reduces the need to place artificial constraints on the players, allows results to account for a wide variety of special circumstances, and facilitates the easy integration of the latest information and interpretation of data without the need for massive updates of data bases and models.

Unfortunately, the approach also has some shortcomings. Because it uses a small team of human experts, the technique requires a significant amount of time to assess large numbers of interactions. The results of the assessments are essentially not reproducible, nor may they be consistent over the course of even a single game. The "model" it uses is difficult to validate because it is not merely a collection of mathematical formulas, but also contains that critical subjective element of assessor expertise, experience, and instinct.

220   The biggest concern about the free method of assessment may be this tendency to produce results of uneven and inconsistent detail, sometimes apparently incommensurate with the importance of an operation (simple interactions being much easier to resolve in such detail than massive battles). Because of the natural tendency to equate level of detail with degree of importance, such effects could mislead players and possibly skew their decisions and the overall outcomes of the game in ways difficult to perceive. Fortunately, the damage assessors at the Naval War College correctly perceive BDA as an input to the game and not an output. Assessors, better than any of the other game participants, know that the results of the game should not be measured by BDA box scores.

Unfortunately, the fact is that most players of professional wargames must measure the success or failure of their decisions precisely by such box scores, or at least by the broad results of the BDA. In this way, BDA propagates through the game both in the effects it has on subsequent decisions the players will make, and also on the players' attitudes toward the "realism," and so the usefulness, of their game experience.

From the inside, the War College method is quite reasonable. The players, however, almost never see things from the inside. Instead, they must base their opinions and decisions on the results of BDA, the basis of which they too often do not understand. The problem seems to center around the difficulty players have in distinguishing unexpected results from unrealistic ones In some cases, the players simply do not understand the capabilities of the systems or forces involved; in others, an "expected value" mindset may make it difficult to accept results that are drawn from the tails of the distribution.

Finally, players must be informed about the results of interactions. Most hobby board games are open games. They provide players full information. Hobby computer games and many professional games limit a player's knowledge of the results of interactions in ways consistent with his ability to learn about the actual situation. If there are no sensors available to determine the effectiveness of a long-range missile strike, the players should receive no information. If the main sources of damage reports for an aircraft strike on an enemy surface force are aircrew debriefs, the potential inaccuracies of such debriefs should be specified in the game rules and applied by the umpires. Time delays for the acquisition, interpretation, and communication of intelligence reports should be similarly specified in the game's rules.

Because actual military operations are often plagued by "the fog of war," professional wargames generally attempt to be at least partially closed in nature. Unfortunately, closed-game designs place heavy demands on the control group. To restrict player access to information, a filter must be placed between them and "ground truth," or game reality. The control group must screen and modify information to delay, confuse, and in some cases misrepresent what is actually happening in a manner approximating what players might experience in actual combat. Such games require large, well-trained, experienced control staffs, such as that at the Naval War College. The control staff must be well practiced and disciplined at providing the quality and quantity of information the players deserve. In addition, players must be physically separated in a closed game, requiring greater space and more extensive facilities in general.

Despite the difficulties of managing a closed game, there are many important issues that simply cannot be explored with open gaming. Obviously, there is little insight to be gained about the roles of command, control, communications, and intelligence (C3I) from a game in which players are given complete and perfect information.

Sometimes, however, good game design and good game players can overcome some of the problems of open games. These latter are often easier to execute, place fewer demands on the control group, and cause less frustration to players. Although the uncertainties of closed games are eliminated, open games are by no means useless. Indeed, for many training and exploratory functions, open games can work quite well, so long as their artificialities are kept in mind. In the end, the trade-off between open and closed games may be dictated by the limitations of facilities and people, and the designer must find a way to limit the negative influences such necessary restrictions may have on the games' effectiveness.

System or Seminar Games

Although it would seem that system games, especially closed system games, might be most appropriate in the defense arena,

222 seminar games do have certain important advantages that assure
their long-term popularity. First, seminar games allow a freer in-
terchange of ideas among participants. Thus, in many ways,
seminar games may be better suited to some training and educational goals than system games. For the same reasons, seminar
games tend to be more effective at helping groups of people arrive
at a consensus about the desirability or feasibility of certain
courses of action.

System games are more rigid and perhaps in some ways
more realistic than seminar games. However, there is a danger in
system games that any departures from reality, due to mistakes in
modeling or input values, or simply outdated structures, can be
more difficult to detect and adjust than in the more open forum
of a seminar game. Balancing the rigor of a system game with the
flexibility of a seminar game to achieve the game's objectives is
thus a critical element of game design art.

The Role of Time

One of the most crucial of game procedures—management of
time—cuts across and affects all of the functions of procedures
and umpires. Time is a critical aspect of any wargame, and the
effective treatment of time is an essential ingredient in any good
wargame design. There are two reasons why this is so. First, in
reality, timing and speed of execution are often decisive in determining the success or failure of a military operation. Second,
time management in a game very often determines the extent of
activity that the game can explore.

In games such as chess, player activity is sequenced in a series of alternating turns or moves. This same terminology is applied to wargames, but in wargames, moves "represent definite
periods of real-world time, . . . [and the] length of a move is the
interval of time for which decisions and evaluations are made."[7]
There are three basic approaches to defining the length of a
move. One approach uses a game clock that runs continuously
and that may be set to run either faster or slower than real time.
The other approaches operate on blocks of time rather than on a
continuous clock.

Strictly speaking, a continuous-time game does not really
have moves in the sense that term generally implies. In such systems, like the one used by the Naval Warfare Gaming System,

players give orders, and forces attempt to execute those orders continuously. There are many advantages to the continuous approach, especially because it appears to be more faithful to reality and is more likely to produce the kind of dynamic interactions that occur in real operations. The price lies in potential distortions, especially in the planning process, when the game-time to real-time ratio (or game rate) is not one-to-one. If game time is speeded up as is the usual case in operational or strategic games, so that one minute of real time represents several minutes of game time, players may find that realistic planning of operations takes too long in game terms and is replaced by seat-of-the-pants or reactive decision making. At the other extreme, if the game clock is slowed down, as is most likely in tactical-level games, so that players may study a situation more carefully before acting, a false impression of the effects of time pressure may easily result. In large and complex games like the Naval War College's Global War Game, some commands may be more heavily engaged than others, requiring slower clock speeds. This phenomenon can force overall game play to run at the slowest speed, or to fragment into multiple "time-lines," thus making game control more difficult.

Alternatives to the continuous-time approach define the amount of game time each move represents and execute the move as a block of time rather than in a continuous fashion. For example, a game may define a single move to represent an entire day's operations; the sequence of events within the day would then be evaluated by umpires or some other system device. Most manual wargames, whether hobby or professional, have used this approach.

There are two fundamental approaches to incremental time games. The first approach defines "the smallest practical period of time . . . and all moves in the game are of that length."[8] This approach is standard in hobby wargames, in which a game turn may represent any span from a few seconds (in a game of tactical air combat) to three months (in a strategic-level World War II game). This fixed-span approach is usually most successful when the span of time each move represents corresponds to the amount of real time the player roles would normally require to collect information, interpret it, make decisions, and implement those decisions.

224   The second approach to time-increment moves is more flexible; moves may represent varying amounts of real time depending on the importance and intensity of activity expected during a given span. For example, a pre-hostilities move may encompass ten to fifteen days of activity. The D-day move of the same game may represent just a single day. Variable-length moves may be predetermined in the game's design or dynamically defined during game play. In the former case, the designer may use the scenario, likely player actions, and play testing to decide on the length of each move. In the latter case, the game director "estimates the time of the next critical event and calls for a move of corresponding length."[9] In some situations, the second approach can be used even in fixed-move games. If moves are specified to be one day in duration, but control personnel believe that there will be little activity for several days, multiple moves may be conducted as a block. Care must be exercised, of course, to prevent distortions whenever play goes beyond the game's intended design in this manner.

Finally, there are two approaches for dealing with the amount of real time players expend in making a move. Most navy games predefine the time allocated to each move in order to assure the completion of all planned moves. For example, two moves per day, one in the morning and one in the afternoon, is a fairly standard scheme. The alternative, again frequent in hobby gaming, allows each move to proceed at its own pace, some going quickly and others moving slowly. Not surprisingly, fixed time allocated for moves is often associated with variable-length moves; free time is often associated with fixed-length moves.

All of these various approaches to event sequencing and time management seek to balance the need to play the game expeditiously, both to prevent player boredom and to explore slow-developing issues, and the need to give players enough time to plan their actions and prevent unrealistic time constraints. The choice of approach to defining moves and allocating time to play them is a critical design decision. As with most such decisions, a judicious use of all the available approaches is often necessary.

Document the Design

Those who have only participated in or observed the play of a wargame may be excused for believing that the most difficult part of game design is creating the mechanics that tie the data and decisions together to let the players reproduce a version of reality. Game designers know better. Although the design of the game system is often the most creative and intellectually challenging phase of the game design process, for pure hard work nothing beats documenting that design, especially writing the rules. This is probably more the case for hobby-game designers than for their professional counterparts.

The quantity and quality of documentation of professional game designs varies widely. In part this is the result of a surprising lack of standards in the field. In part it is a result of the fact that few professional wargames are designed by a single individual, as is often the case in the hobby, and even fewer are built completely from scratch. Instead, the various components of the game (scenario, data base, models) are documented separately, by the different individuals or teams responsible for their creation. Unfortunately, the step that is almost never documented is that final act of picking the pieces to be used and devising how they will all fit together. Indeed, much of what the professional community calls game design can revolve around adapting scenarios and models to the long-standing procedures of the particular organization responsible for carrying out the game. In too many cases, these procedures are passed on from generation to generation with little written documentation (or at least little documentation of recent practices).

Such was the case at the Naval War College from the later 1960s, when Frank McHugh published his major works on gaming. A big step forward was made in the mid-1980s when the college's War Gaming Department published some important papers on how to perform battle-damage assessment and how to play the opposition side in War College games. The designs for individual games, however, continue to be documented only by a relatively short paper that outlines the game's objectives, overall structure of play, and administrative schedules of events. From that point on, it is the control staff that makes the game go; a solid and experienced team can produce a useful and informative game with only minimal actual design effort by the designer.

In the world of hobby games, experienced control staffs cannot be packaged in every box. Instead, the designer must craft well-structured, reasonably complete, and easily understood (at

226 least when compared to a textbook in quantum physics) rules of play that people inexperienced with the game system, or even with gaming in general, can use to play the game correctly. It is no easy task, as the existence of a mountain of game errata for nearly every hobby game ever published attests.

There are different ways to write rules for different communities within the hobby. Although improving, rules for computer games—the new kid on the block—are often the least rigorous, complete, and detailed because the machine itself can act in many ways like an experienced and unfailingly consistent umpire. Players can be told how to implement their decisions without telling them how to resolve the outcomes of those decisions. By taking the calculation of combat results (or battle-damage assessment function) away from the players and giving it to the computer, such games begin to approach the feel of a professional wargame.

Miniatures games, the grand old man of the hobby, must have more complete and detailed rules than computer games, but they are, in fact, seldom provided in the published rules themselves. Miniature game rules have a long tradition of looseness (or flexibility, as miniatures aficionados prefer to call it) and miniature gamers an equally long tradition of devising "house rules" and agreed-upon interpretations of printed rules. Indeed, many miniatures gamers begin with a set of published rules and so modify them that a newcomer to the group would find the original rules unrecognizable. The process of writing and rewriting rules that H.G. Wells describes so eloquently in *Little Wars* is still accurate for today's miniatures game designers.

It is in board-game design, however, that we find the most rigorously demanding task for the rules writer. The board-game designer cannot rely on the *deus ex machina* of the computer, nor on the tradition of debate and compromise supplemented by a human umpire used in miniatures gaming. In a board game, the designer's cards must literally be placed on the table. Everything the players need to set up and play the game must be provided by the components and rules. Not only must the players be told what they can do, they must also be told precisely how to do it and how to resolve the outcomes.

The complexity of board-game rules naturally grows with the complexity of the situation the game portrays. Early board games like Avalon Hill's *Tactics II* and *Midway* required only on **227** the order of four 8 1/2″ × 11″ pages of basic rules and perhaps an equal amount of advanced and optional rules. More recent games such as Victory Game's *7th Fleet: Modern Naval Combat in the Far East* have increased the number of pages devoted to game rules by an order of magnitude.

Over the years many styles of writing rules have developed, some associated with particular designers (such as the conversational style of a John Hill), and some with particular publishers (such as the legalistic style that became synonymous with Simulations Publications, Incorporated in the 1970s). The original Avalon Hill style was fairly loose, but accurate enough for the relatively simple games of the early 1960s. Simulations Publications led the way into the realms of greater complexity with a rigorous "case system" structure and a dry, formal style of presentation that sought to master the complexity of their rules by airtight legislation. Unfortunately, this rigid style of rules presentation often made it difficult or impossible for players to understand the rules as a whole. Reactions to the Simulations Publications approach led Game Designers Workshop and the later Victory Games group to a more balanced blend of tight structure and less-formal wording, designed to allow the players to read more easily and understand the rules while at the same time covering the multitude of situations that could arise during play.

Ultimately, no matter what the format of the game or the style of presentation, good hobby-game rules are based on some fundamental, and fairly obvious, principles:
- Adapt the rules to the game, and not the game to the rules.
- Tell the players everything they need to know to play the game by structuring the rules around the sequence of play.
- Provide plenty of examples to illustrate how the rules are supposed to work, both individually and in concert.
- Explain the underlying rationale for particularly important or especially unusual rules.
- Integrate the text explaining the rules with the graphical play aids designed to help implement them.

Nothing very surprising there, yet unfortunately the professional gamers have for the most part failed to adapt these basic ideas to their games. In too many professional games, the players are given no idea of the rules at all. Certainly, the player in com-

228   mand of a fleet has little need to know the computer code that implements a model of air-to-air combat, but he could find it helpful to know what characteristics of a strike force make it more or less vulnerable to enemy interceptors. Once again, the professional wargamer needs to look to the hobby to find a more balanced approach to documenting the design of the game and providing the key elements of that documentation to the players so that they can improve their play, and their learning experience during the process.

6

DEVELOPING
WARGAMES

WHAT IS WARGAME DEVELOPMENT?

The word *development* when applied to wargames is another example of that unfortunate lack of precision that has plagued the terminology of the field from its inception. If they thought about it at all, most people unfamiliar with the board-wargaming hobby would probably assume that "game development" refers to the entire process of producing a game, from the initial gleam in the designer's eye to marketing the finished game. Board gamers, on the other hand, associate the term with a specific major step in that production process, a step that the hobby-game publisher ignores or shortchanges at his peril, but one that is all too often neglected by professional gamers.

The application of the word *development* to a specific phase in the production of a game can be traced back to Jim Dunnigan's tenure at Simulations Publications, Incorporated. In issue number four of *Moves* magazine, Dunnigan and his team described the development stage as the one in which most innovations came about.[1] The process involved one or more "developers" and play testers. The designer, who had previously integrated his research and the fundamental concepts of the game system, presented the skeleton of the game to the developers, whose responsibility was basically to turn that skeleton into a fully fleshed-out and, most importantly, playable game.

Working with the designer (hopefully!) the development team first debugged the game, making sure that it actually could do what the designer said it would. After a round of play testing, the team prepared a revised rules outline based on the designer's first draft or rough notes. As additional clarifications and explanations (and sometimes entirely new rules) were incorporated, the outline was expanded and put in final form during another series of play tests. Finally, the complete rules were drafted and edited, the components sent off to the art department, and the final production phase began. Thus, in the Dunnigan/SPI system, the roles of the game designer and developer often overlapped (sometimes too much so!), particularly during the documentation of the game's design.

Prior to Dunnigan's institutionalization of the game development process, the testing and refining of a game design was pretty much a hit-or-miss affair. Some games, like Dunnigan's own designs of *1914* and *Jutland*, were very well designed, but lacked a real development stage. As a result, they were, to many unsuspecting purchasers, quite difficult or impossible to play. (As an aside, my doctoral thesis adviser, a Ph.D. statistician, gave up on *1914* as too complex!) The earlier Avalon Hill games, based largely on the fundamental systems introduced by Charles Roberts in the late 1950s, had, in a sense, been developed by the experience of their predecessors. As newer games and more complicated concepts were introduced, the development stage became much more specific, and critical.

Yet, if Dunnigan can claim credit for the name, the importance of game development was much earlier and more vividly described, once again, by Mr. H. G. Wells. *Little Wars* gives a thorough description of the inception, initial design, and subsequent development of Wells's game system.

As Wells described it, the game was born out of one of those happy accidents when, after lunching in a room "littered with the irrepressible debris of a small boy's pleasures,"[2] an acquaintance of Wells, referred to only as Mr. J.K.J., used a breech-loaded, spring-powered toy cannon to fire a small wooden cylinder and knock over a toy soldier. He challenged his friends to try their prowess, and by the end of the session someone had proposed the crucial idea: "'But suppose,' said his antagonists, 'suppose somehow one could move the men!' and therewith opened a new world of belligerence." [3]

Wells soon built on this basic idea. He and his friends began using volumes of the *British Encyclopedia* to construct a "Country" in which to maneuver their forces of infantry, cavalry, and guns. They also devised a basic system of alternating moves and hand-to-hand combat. The game design, as we would call it today, was complete, but it had certain problems.

The hand-to-hand system used the flip of a coin to determine the outcome, allowing occasions on which "one impossible paladin slew in succession nine men and turned defeat to victory, to the extreme exasperation of the strategist [dare one assume Mr. Wells?] who had led those victims to their doom." [4] This heavy effect of luck was combined with allowing the guns so much freedom to move and fire that battles turned into "scandals of crouching concealment," with the guns popping away at the troops cowering behind their books.

Enter the development stage. In a flurry of creative activity, Wells solved the bulk of the problems. The terrain was improved by using cardboard and children's blocks to construct houses. The power of the gun was reduced by allowing it either to move or fire (but not both) in a single turn, and by demanding the presence of soldiers nearby in order for it to be fired or moved at all. Another series of games was played using these rules, and new ideas were introduced, this time with the professional assistance of "Captain M., hot from the Great War in South Africa." [5] Through this series of rules proposals and play testing, Wells's game finally reached the final form, whose rules he presented in *Little Wars*. A better description of the process of game development is nowhere to be found.

At the risk of lapsing into cliché, then, we can summarize the relationship of game design and game development in a simple phrase: design proposes and development disposes. Just as any article, paper, or book will benefit from a thorough job of editing before publication, even a well-designed wargame will benefit from thorough development. Although the process may be formal or informal, game development seeks to ensure that the game design is complete, as realistic as possible or desirable, and that it is playable and capable of meeting the objectives specified for it.

232 THE GOALS AND ACTIVITIES OF GAME DEVELOPMENT

A formal summary of the goals of the development process and the principal activities involved in achieving these goals are listed below.

Goals and Activities of Wargame Development

Goals ensure that:
- The pieces of the game do what the designer intends
- The necessary pieces are available
- The pieces fit together
- The entire game functions smoothly
- The sponsor's objectives can be met
- The game responds to expected actions in expected ways
- The game can deal with unexpected actions efficiently

Phases of development include:
- Model, data, and scenario validation
- Play testing
- Preplay
- Preparing the final rules

The first goal is to verify that the individual elements of the game actually do what the designer expected and intended. Obviously this requires that the developer understand the designer's intentions. In some cases the game designer may be responsible for the development of his own game, thus reducing the problem of clearly communicating design intentions to the developer. A separate developer does tend to force the designer to be more specific and explicit in defining his expectations and reduces the ever-present danger that unstated assumptions, already internalized by the game's designer, will color the development and testing of the game and make it difficult to perceive fundamental problems.

The need to verify intentions and expectations is perhaps most pronounced in the case of any mathematical data and combat models used in the game. The accuracy of the data and the performance of the models must be checked carefully to ensure that they reflect the designer's research and are actually operating in accordance with the structures he has specified. When such data or models are complex or computerized, this verification process can be an involved one. Programs must be debugged and many test cases run. Similar checks should also be carried out to **233** test for the internal consistency of scenario assumptions and, in the case of hobby games at least, victory conditions.

Another major goal of development is to try to ensure that the game design is complete. All the data and models likely to be needed for play must actually be available. Even more difficult to verify, however, is whether procedures or rules exist to cover all the possible situations that might arise during game play. In professional games, all the foreseeable political or higher-level military decisions that the control team may be called upon to make should be specified along with general guidance to cover unforeseen circumstances. Despite the best efforts of development, however, it should not be surprising that at some time a situation will arise that is not covered by the data, models, or instructions to the control team. All development can hope to do is to reduce such occasions to the barest minimum.

After making sure that the game's pieces work as the designer intended (and that those intentions are reasonable), and after satisfying themselves that all the needed pieces are available, the game's developers must then test the game to be sure that all the pieces mesh into a workable system. A mathematical model of engagements between combat-air-patrol and deck-launched interceptors from an aircraft carrier and attacking bombers armed with air-to-surface missiles might produce the number of enemy missiles that penetrate into the surface-to-air missile defense zone. But if the model dealing with the effectiveness of that part of the defense also requires the time- and altitude-separation of the incoming missiles in order to calculate results, the pieces do not fit together.

In addition to making sure that the various parts of the game system mesh, the developers should also attempt to polish the game's mechanics so that the entire system functions smoothly and efficiently. In a seminar game held by the Naval War College, for example, players may be required to submit written instructions, both general and specific, so that the control team can accurately execute and assess the results of a given move. The game's designer may specify that the "move sheet" contain a section outlining general intentions and also a listing of the major units under the player's control and their movement and mission orders. The game developers may add more detailed require-

234 ments, such as the depth of submarines or the altitudes of aircraft, to the move sheets so that the orders may be more precisely evaluated using the models available for the game.

But perhaps the most important and difficult task of game development is to ensure that the game can meet its objectives. This is especially important for professional games, whose objectives tend to be more specific than those of commercial games. In addition, the distinction between the sponsor of a professional game and the designer of the game may cause some disconnects in the translation of objectives into game design. Although designing the game to meet the objectives may seem an obvious goal and one unlikely to elude the game's designer, it is not impossible for the design process itself to obscure the purposes for which the project was originally undertaken. The natural tendency, and often the requirement, for the designer to spend much of his time working on the details of the game system rather than on the broad outlines of what it should accomplish is another strong reason for assigning someone other than the designer to take primary responsibility for game development.

Finally, game development must exercise the game to see how the system reacts to player decisions. Especially important is the system's response to unusual, unexpected, or extreme decisions players may make. This is especially important for system games, particularly those controlled by computer, in which the stabilizing influence of a human umpire is missing. The player that "cracks the code" of the game may play the game very successfully, but the educational value of his experience and the operational insights to be obtained from his decision making may be nonexistent because he is only taking advantage of a loophole that exists in the game system but would not exist in reality.

Such effects are noticeable even in professional games. A veteran of several Global War Games will have a pretty good idea of what the Red Team will consider to be allowable strategies and tactics, and how Control is likely to respond to certain suggested operations. In the hobby world, loopholes can be much more dramatic. In a recent review of SSG's computer game *Russia*, the reviewer revealed just such a loophole.[6] The game, a technical marvel dealing with various campaigns and the entire war between Germany and the Soviets from 1941 to 1945, uses an artificial-intelligence-type system to allow the computer to play against a human opponent. Unfortunately, the reviewer found that "artificial stupidity is a far better expression for watching the computer constantly press forward in useless attacks."[7] By taking advantage of this quirk, the wily human player can set traps no other human would ever fall into but which the computer opponent embraces as heaven-sent opportunities for victory. Despite its other successful, interesting, and educational features, such a loophole leaves *Russia* an unfortunately flawed game.

As a result of the dangers lurking behind such hidden loopholes, the game's developers must not only try to play the game as the designer expected or wanted it to be played, but they must also try to break the game. Expected decisions should cause the game system to produce the expected responses (although not necessarily the expected outcomes from stochastic models). On the other hand, unexpected player decisions should not cause the game system to self-destruct.

To achieve their objectives, game developers must carry out three related but distinct activities: data, model, and scenario validation; play testing; and preplay, or in the case of a hobby game, blind testing. Although in some sense these activities are performed in the sequence listed, validation and play testing, as described earlier in relation to *Little Wars*, are continuous and iterative processes.

VALIDATION AND GAME "REALISM"

One of the most important jobs of wargame developers is to assess the validity of the game's results and processes in light of the real world. Presumably the designer has tried to produce as realistic a game as possible, given the design decisions and trade-offs he chose to make. The developer's effort, by looking at the designer's product from a more objective distance, can help to judge how successful those design decisions were. In order to do so, however, it is necessary to understand what the terms *results* and *validity* mean when applied to wargames, and also to develop an appreciation for the uncertainties that underlie the results.

There are two types of game results of most interest. The most obvious, and potentially most misleading, results are the game events themselves, the outcomes of engagements, battles, and campaigns, or most particularly in the case of hobby games, the overall winner and loser of the game. Interest in these kinds

236 of "outcomes" centers around who won and how. The second types of results, much more meaningful to professional gamers or to hobbyists more interested in understanding than victory alone, are the insights and issues that arise from the play of the game. Interest in these types of results focuses on why the players made important decisions.

The validity of a game's results can best be thought of as the extent to which those results reflect reality as opposed to the artificiality of the gaming environment. This validity depends on the accuracy of the mathematical evaluations of operational capabilities used in the models of the game, and on how well the quantity and quality of information available to the players of the game reflect the levels of information likely to be available to their real-life counterparts.

The uncertainties that arise during the play of a game are similarly of two types, and are closely related to the two types of results. Stochastic uncertainties arise from the variations in the outcomes of similar operations or engagements that come about as a result of the probabilistic nature of some kinds of events (such as the probability that a torpedo will strike and sink its intended target once it is launched). Such uncertainties are often characterized as the "roll of the dice," and in hobby games are usually the outcomes of precisely that. Strategic uncertainties, on the other hand, revolve around the choice of options open to the players. For example, in a modern naval wargame, there may be quite some amount of uncertainty about whether the Soviets will actually commit submarines to attack the sea lines of communication across the Atlantic or Pacific. Although fundamentally different, these types of uncertainties usually affect each other in profound ways; uncertainties over capabilities can affect the choices of options that, in turn, affect the occurrence and nature of engagements during the game.

One of the first things the developers must do is to determine whether the course of game play and the experiences and insights the players may derive from it are dictated primarily by the game's scenarios, assumptions, and mathematical combat models, or by the decisions that the players are capable of making. When the former dominates the latter, misconceptions and errors in understanding can result. These are especially dangerous problems in the world of professional gaming.

For example, a series of games was conducted in the early 1980s to allow senior civilian and military decision makers an opportunity to explore how the U.S. Navy and Marine Corps could best act to deter war in a crisis, and to defeat the Soviet Union if the crisis degenerated into a war. Although this question is central to many navy games, this particular series had a smaller and more select body of participants than the Global War Game series, for example. Unfortunately, the conduct of the games fell into an all-too-familiar trap. The strategic questions were already assumed to be answered, for the most part, before play began. Player attention during the game focused principally on lower-level operational issues, such as which aircraft carrier was deployed where, and how many submarines were killed on a daily basis. The decisions of the game players, who represented the National Command Authority and various commanders in chief, seemed to have little effect on the course and outcome of game events.

In essence, these games used senior officers and officials to make "decisions" already preordained by scenario imperatives and assumptions of military capability. This type of practice can lend unwarranted credence to concepts of operations and models and estimates of systems effectiveness that have not, in fact, received the thorough scrutiny and approval of professional judgment that their use in such a game may imply. This scrutiny is the developer's primary responsibility.

But how is a developer to determine if a single mathematical model is realistic, much less a large and interacting collection of such models? The problem may be made even more difficult in a professional game (or even a hobby game) dealing with a projected, rather than historical, scenario whose own realism or even reasonableness may also be difficult to judge.

The debate about the validity and utility of mathematical models of warfare is probably as old as the practice of devising them. In the professional world, opinion ranges from one extreme to the other. For some, the attempt to build mathematical models of warfare, especially of future warfare, is hopelessly doomed. The modeler's "attempts at analytical rigor are not empirical science but *a priori* modeling. They rest on analysts' judgments as to what criteria are pertinent, which parameters are critical, how the parameters interact, and what values they will

238 take. . . . Or as Keynes said of mathematical economics, warfare models 'are mere concoctions, as imprecise as the initial assumptions they rest on, which allow the author to lose sight of the complexities and interdependencies of the real world in the maze of pretentious and unhelpful symbols.'"[8] This attitude basically reflects a belief that "it is impossible to quantify warfare without having a war to quantify. Only then are there data embodying the true complexities of combat."[9]

Others, who believe in the utility of mathematical models of warfare under at least some circumstances, reject attempts at "the quantification of psychological or human military qualities which are unquantifiable."[10] Those who hold this position believe modelers should simply ignore "psychological dimensions: their uncertainty, their variety, their inconsistency and their lack of utility in modeling future conflict should make one reluctant to expend resources pursuing them, leaving the community the time and money to quantify better the quantifiable."[11]

To others, however, mathematical models and analysis are necessary and useful tools, but only part of the equation: "'Descriptive' models enhance our understanding. 'Predictive' models help describe temporal, spatial, and organizational relationships."[12] But modelers must temper their work with an understanding of history and technology.

The real dangers in modeling lie in the "inability to cope with uncertainty, and to place confidence limits on . . . findings. . . . It means that the battle models . . . are broken reeds, a totally inadequate basis for the advice operations analysts provide their military clients."[13] Analysts holding this view also must accept the fact that until modelers "tackle the problem of bounding both the uncertainties in the inputs and their effects on the outputs there is no salvation in improving models in other respects."[14] Such may not be an easy task.

Physicists had to deal with the same source of difficulties in the process of reconciling the various characteristics of subatomic particles. "But if the models of physics, with all of their power of precision and experimentation are confounded by inherent uncertainty, and if as some physicists say, quantum mechanics cannot be explained except by analogy, then surely [operations analysts] (and our critics) should be charitable in our demands on our power to measure the quantities of war. In particular, where is the research that will tell us the residual uncertainties in measuring (never mind predicting) those quantities?"[15]

In time of war, as was the case during the birth of operations research during World War II, actual data might be available to conduct such research. During peacetime, the only potentially useful sources of data, aside from infrequent and generally small-scale low-order conflicts, come from history. Indeed, in a recent study conducted for the U.S. Army, over 260 modern battles were investigated in an attempt "to provide a tool for judging whether the results of simulated combat are consistent with historical combat."[16] Although only a start, such studies hold some promise of producing some of the basic research necessary to put combat modeling on a sounder footing.

But data alone are only a start. In the wargaming hobby, most games deal precisely with past, historical combat. Even when data is available about such combat, translating that data into an accurate or "realistic" model of combat requires more than a little dose of experience, cleverness, and talent. Although the quality and quantity of data varies, there is usually at least some basis in fact on which the models of historical games can be based. Even more importantly, there is the historical record against which the results of the game system can be checked. It is in the interpretation of the available information, and in its adaptation to the particular game environment, that differences of opinion and of style may arise.

One of the hobby's most respected designers, Frank Chadwick, wrote that "there are two distinct forms of accuracy: product accuracy and process accuracy. The former, if viewed in complete isolation from the latter, is typified by the 'it all comes out in the wash' school of thought. . . . The latter, again in isolation, holds that the final outcome is irrelevant; what's important is the feel of each turn."[17] Although most hobby designs obviously must incorporate elements of both philosophies, some designers emphasize one approach to accuracy far more than the other.

The champion of the design-for-the-right-feel school is John E. Hill. Hill's attitude is summarized in his belief that the "main problem with the whole field of game design philosophy is that it is operating under the mistaken belief that it might really be possible to simulate the chaos of war with cardboard and paper."[18] This attitude is at the foundation of Hill's "design-for-effect" philosophy, a philosophy whose proponents "are not as interested in 'what actually happened' as they are in the effect of that event."[19] The result is that a game becomes less a simulation of reality and more a simulation of the designer's interpretation of reality.

In his unpublished designer's notes to the game *Napoleon's Last Triumph*, William Haggart articulates the opposite viewpoint. For a "design to be 'accurate,' a true model of historical reality, two points must be demonstrated:

1. There exists a corresponding fact substantiating every cause-and-effect relationship. . . .
2. The historical 'facts' presented as a basis for the accuracy of the design must be verified, proven."[20]

Of course, as Haggart realizes, merely researching the historical "facts and interpreting them does not necessarily generate an accurate simulation, even when done perfectly. The simulation still must duplicate the historical data in play to be termed 'accurate.' After completing the design, [he] used the following process to verify the game's integrity:

1. Each subsystem was individually played to ascertain whether all the possible combinations and results would remain within the historical limits set by the data.
2. The same testing was done by pairing two, then more subsystems . . . to determine the parameters.
3. The total game was tested against the actual events of the battle: could the game re-create the original progression of [the battle] within the rules? Were the casualties, combat, and leader performances close to the real thing?

Changes were made at every level until the game system played accurately and effortlessly, without a lot of 'special rules' and exceptions."[21]

Hill's philosophy and Haggart's philosophy are as different as "a painting and a photograph of a battle. The painter looks at the battle, and decides which aspects he thinks are most important, either because they interest him more or, if he is trying to give you a close picture of reality, because they seem to him the essence of the scene. . . . The photograph shows precisely what the battle scene looked like at one moment in time. The obvious influence of the painter is eliminated, but the photographer can insert his view nonetheless, by selecting one photograph from many. . . . In the same way, game designers pick the aspects of battles which they feel were important to the outcome. No game can include every factor: the designer must choose." [22]

It is in making this choice that the designer's art and philosophy are tested, and in evaluating the validity of the choice that the developer plays his most important role. Professional operations researchers and systems analysts debate the question of model "validation" in endless and often fruitless exchanges. In only the rarest of cases is any real sort of validation possible, when actual present-day combat data is available to compare with model results. The war between Great Britain and Argentina over the Falklands/Malvinas Islands provided a limited quantity of such data, which was used, for example, by Mark Herman to "validate" the Tactical Analysis Module of the Strategic Analyses Simulation. The analysis demonstrated that the model could, indeed, produce results consistent with what occurred in reality. In order to do so, however, the "verniers" or scaling factors of the model had to be adjusted correctly.

It is quite impossible, in the hobby world or the professional world, to build a wargame or a combat model that is *certain* to reflect accurately the reality of future combat for the simple reason that we do not *know* what that reality will be. But that is not the point.

The only realizable goal for a model of future warfare is to reflect, in the most complete and coherent way possible, the analysts' (or the analytical community's) beliefs and understanding of the key elements of that combat. By exercising, testing, and modifying that model, analysts and wargamers can explore the implications, not of some unknowable future reality, but of our current, restricted, and uncertain view of what that reality might be like. We can do no better than to try to identify the hidden interconnections and consistencies of our current thinking as objectively as possible. But such a goal, as limited as it may appear to those who seek crystal balls in computer code, is not only a worthy one but essential as well.

What is crucial for wargaming, with its focus on creating as accurate a decision-making environment as possible, is that the

242 models reflect our best-available understanding of the factors and conditions that affect the player's decision-making process and his ability to gather and interpret information. Fabricating such an accurate environment is easier for some levels and types of warfare than for others.

In hobby wargaming, for example, there is a strong body of opinion "that certain levels of games (notably true tactical games and grand strategic games) may by their very nature defy treatment as accurate as is the norm for games in the operational vein. . . . The fundamental difference between [the operational level] and both the true tactical and strategic level is that at those levels the uncertainties in decision-making have to do with, to use a game analogy, what the rules are."[23] Yet, some of the effects of such uncertainties can be factored into the games through the use of umpires or randomization of those aspects of play beyond the influence of the player's real-life counterpart.

The key to creating accurate models and systems, then, lies in the ability of the design to incorporate as many of the elements important to actual decision makers as possible while simultaneously minimizing the number and effect of extraneous factors. Only by trying out the models and their interplay and evaluating their ability to create as accurate a decision-making environment as possible, both in its "feel" and in its intellectual scope for command, can the developer "validate" a wargame's "accuracy."

PLAYTESTING

Although professional gamers are sensitive to both the need to validate and the difficulty of validating models, they are often not quite as sensitive to the requirement to validate the entire game system. This perception contributes to the lack of serious playtesting of professional games before the game is played "for real." (To be fair, the constraints of time play another and often dominant role in the shortage of playtesting of professional games. Indeed, as described later, some games are tested during a preplay phase.)

In the hobby world, on the other hand, playtesting is a familiar concept. It is, in fact, the principal tool of game development in addition to being one of the process's major phases. As the word implies, playtesting combines playing the game with thoroughly testing its functions and its ability to meet its **243** objectives.

It should not be surprising that the best way to determine whether all or part of a game system works is to play it. What may not be so obvious is that playing to test is significantly different from playing to enjoy or playing to learn. In the latter cases the players are usually working with the game, trying as much as possible to cooperate with the demands of the system. Testing is much more of an adversarial relationship; the tester should never give the game the benefit of the doubt. He should try his very best to make the game fall apart. Only by doing a thorough job of trying out the various alternatives provided by the game's design can the testers be sure that everything works as intended.

But playtesting is not just a negative process. Playtesting can often suggest workable solutions to the problems it discovers. It can also point out the importance of aspects of a situation that the designer overlooked or chose to ignore. Just as importantly, it can often suggest easier to understand, more efficient, or more exciting ways of carrying out the functions of the game than those originally conceived by the designer. Professional gamers sometimes seem to have the attitude that ease of play is only of minor importance in their games. Yet the fact remains that the more efficient a game's mechanics are, and the less attention the players must pay to the overhead of gaming artificialities, the greater the chance the players will become more involved in the simulated reality of the game. Excitement flows from this involvement, from the challenge of the situation, and from the pace of activity. Finally, involvement and excitement produce the atmosphere of intense competition that can lead to the development of original ideas so important to the value of professional gaming.

The professional community should take a leaf from the hobby book on the subject of playtesting. Playtesting a professional game should involve playing the game as the designer intends it to be played, but without the actual participants. Instead, playtesters should assume the roles of those participants and attempt to test the functioning and the limits of the game. The integration of the various game elements can then be exercised and refined. Not only must the playtesters check to see if the game works as expected, but they must also try their best to make the

244 game break down. If this process discovers problems with the system, procedures, or data, they can then be corrected before the actual play of the game.

The importance of the testing aspect of the playtesting phase cannot be overemphasized. Thorough testing cannot guarantee a problem-free game, but insufficient testing almost certainly increases the chances of having to cope with unforeseen difficulties during play. Unfortunately, thorough playtesting requires the allocation of more time to the entire game preparation cycle and the commitment of more people to the playtesting process.

PREPLAY AND BLIND TESTING

The last step in the game development process is a kind of dress rehearsal for the game. In professional wargaming this step is usually referred to as preplay. The equivalent hobby concept is usually known as blind testing.

At the Naval War College, preplay focuses on the mechanics of game control and the availability of required information.[24] Game-control staff members play through the mechanics of the game in an abbreviated fashion to familiarize themselves with their responsibilities and to prepare themselves to handle some of the situations they may be called upon to address during the upcoming play of the game.

Because of the great demands of thorough playtesting, it is sometimes necessary to combine playtesting and game preplay. As practiced at the Naval War College, preplay seeks not only "to check the NWGS computer play file of information, to validate [the] data base and, most importantly, to prepare the WGD control group personnel . . . for the upcoming game . . . [but also] to expose difficulties or omissions in preparation prior to the arrival of the players."[25] The War College has been remarkably successful with this compressed approach, but the difficulties of adequately carrying out both testing and preparation often leave the testing function to take a back seat.

In the hobby, blind testing may be considered "nothing more than having the game played without the designer or developer in attendance." The intention is to give the testers "the closest thing to a finished version of the game" in the hopes they will play it as if they had just purchased it in a store.[26] This is the final check, looking for the hidden bugs that designer, developer, and experienced testers may have missed because the game had become so familiar to them.

PREPARING THE FINAL RULES

Throughout the development process, the designer's original set of rules constantly changes and evolves toward the finished product. Errors are discovered and fixes are found. Holes are filled and wording made more precise. Just as the designer documents his work in the draft of the rules turned over to the development team, the designer and developers together document the results of the development process during the preparation of the final set of rules. All the rough notes, comments, and suggestions of the playtesters must be refined and rewritten into the text of the game's instructions.

A hobby game that has passed through a thorough, systematic development and blind-testing program, and has had the results of that program carefully crafted into its final rules, is as close to error free as a game can get before publication. Yet, it would be surprising indeed if there were not still some hidden and unexpected quirks that no one involved had discovered. The ultimate test, and the ultimate proving ground, is in the playing.

7

PLAYING WARGAMES

ROLE PLAYING

There has been something of a debate in the gaming hobby about whether the players of wargames, particularly of historical board games, should be cast in the roles of specific real-life decision makers. Some of the disagreement is sparked by the existence of what is known as the fantasy-role-playing hobby, represented by its most notorious example, the games known collectively as *Dungeons and Dragons*. The self-styled "real wargamers," who are often aficionados of the study of military history or technical minutiae (such as the ground pressure exerted by the Tiger tank), sometimes feel uncomfortable if they are referred to in the same breath as their "eccentric" cousins whose concerns center on the characteristics and fighting power of elves and goblins (central characters in most fantasy gaming).

But if the essence of wargaming is decisions, then the essence of playing wargames lies precisely in the player's ability to assume a meaningful decision-making role. In other recreational games such as *Monopoly*, chess, or poker, the only role of the participant is as a player in the essentially abstract world of the game; his value system, resources, and decisions reflect only that

248 abstract reality. Such games are often excellent pastimes and diversions; they may even be financially profitable to their players (to some poker players, anyway). A wargame provides players with the opportunity to assume the roles of military or political decision makers in a simulated representation of the real world. To the extent that the game reproduces the environment that the real decision makers must face, it can provide its players not only with diversion, but also education, training, and an increased understanding of the real world.

Well-respected historical-board-game designers such as Frank Chadwick of Game Designers Workshop have argued that "all historical-based games are role-playing games."[1] In the last few years, computer-wargame designers like William Nichols (*Seventh Fleet* and *Long Lance*) have also come to the same conclusions. Nichols's games make use of what he calls the "viewpoint oriented" approach, in which the player's knowledge and control over the situation is restricted to as accurate a representation of the limitations of real-world command as the game's designer can devise.

Counterarguments to the role-playing interpretation tend to focus not on whether players do, should, or must play roles, but rather on whether they should be restricted to only one such role in a given game. In a board game such as *6th Fleet*, the highest level of player decision making is that of the theater or fleet commander, yet the bulk of the playing (and incidentally the fun and interest) of the game takes place at the battle-force commander's level, and even sometimes at the level of the individual ship-driver. Restricting the player to the role of the fleet commander would leave him little to do and reduce the game's enjoyment value tremendously. The price of introducing this additional source of fun, however, is that the player has much greater control over the tactics and detailed dispositions of his subordinate forces than any real-life commander ever enjoyed.

The key point in the debate is not whether players should play well-defined roles (clearly they should) nor how many such roles a single player should have (not many); the key point is whether the structure of the player roles contributes to or is detrimental to the game's realistic representation of the operational situation, and also to the player's willing suspension of disbelief. If the player's role is too limited in its scope for decision and action, the player will lose interest and remember that "it's only a **249** game," and not a very good one at that. If the player is overloaded with detailed decisions that are not properly the responsibility of his principal role, he may lose sight of his primary objectives or get bogged down in the details of the game's modeling.

In the professional wargaming environment, the role-playing aspects of the game are, perhaps surprisingly, more widely recognized and accepted. Part of the reason for these attitudes may lie in the fact that professional gaming originated in the educational and training sphere, where role-assumption is a fundamental feature of gaming's utility. Despite their theoretical understanding of this facet of playing a wargame, however, once players of a professional game arrive at a game site and the game is actually underway, they often have trouble staying in their roles. Why? Typically the problem centers around the second of the difficulties described above—that of getting bogged down in detail.

As discussed earlier, most officers who play in a professional game assume roles of commanders whose real-life ranks are higher than those of the players. As a result the players are often less familiar and less comfortable with their game roles than they would be if they were called upon to perform the tasks involved in the positions they currently occupy or have held in the past. If given the opportunity to become involved in issues at these more familiar levels, most officers will tend to spend much of their time and effort on those lower-order issues rather than on the less-familiar tasks of their principal game role.

The most difficult time for players (and control staff, for that matter) to keep to their roles usually occurs when control reports the results of damage assessment, especially when those results differ significantly from player expectations. If an air strike against enemy shore targets results in extraordinarily high losses to attacking aircraft, the players (especially those who are familiar with the kinds of mathematical models used in the game-evaluation process) too often complain about the quality of the models used to produce the outcome (or sometimes about the lineage of the damage assessor) instead of inquiring about the possible operational explanations for why the losses were so high.

Failures of reconnaissance, poor tactical execution, or misestimation of the enemy are to be expected in warfare and should be reflected in game play as well. So, too, will the extraordinarily

250 favorable circumstances players sometimes overlook when they are pleased with a successful outcome. A well-designed game and well-prepared control staff will seldom produce or report an assessment that is beyond the bounds of reasonable possibility. Unfortunately, sometimes the game control staff contributes to the problem by reporting results in terms of model outputs rather than couching those results in terms of the outcomes of actual combat. Such opportunities to break the spell and allow the players to return too far back to real reality instead of remaining focused on game reality must be avoided at all costs if the players are to continue suspending their too readily resumed disbelief.

LEARNING FROM PLAYING WARGAMES

Because it emphasizes human interaction and role-playing, wargaming can be a powerful learning tool. Participation in a game allows players to "practice" the roles they assume in the game. Because wargames are not real, however, there are significant limitations on the extent and validity of what can be learned by playing them.

Admiral Arleigh A. Burke put his finger on the central artificiality of wargaming when he said that "nobody can actually duplicate the strain that a commander is under in making a decision during combat."[2] In a wargame, real forces do not deploy, real weapons do not engage, and real people do not die. Wargames, like exercises, are only an imperfect image of real war, no matter whether they are the paper and cardboard images of the hobbyist or the sophisticated computer images of the professional. To understand what can be learned from playing a wargame, it is necessary to understand which game experiences are most like what goes on in actual combat operations.

In an actual military operation, a commander is assigned a mission and allocated the forces with which he is expected to carry out that mission. The commander and his staff must plan how they will accomplish the mission, communicate that plan to their subordinates, and then monitor the execution of the plan by their forces. During the planning phase, the commander must analyze his objectives and the alternatives for attaining them, assess the enemy's capabilities and possible courses of action, and identify his own strengths and weaknesses. He must then choose a concept of operations that appears to have the best chance of success. In many cases, he must understand not only military and technical factors, but environmental and political ones as well, translating all of these into operational opportunities and devising ways of exploiting those opportunities to accomplish the assigned mission.

Once the plan is complete, subordinates must be informed of their roles and how they are expected to perform them. The commander must clearly explain his concept and identify those decisions he reserves for himself and those he delegates to subordinates. To control the execution of the plan, the commander must specify what information he needs to make his decisions, how he expects to receive, store, and access that information, and how he plans to evaluate it. Once the operation is underway, the commander must integrate the information he receives with his own tactical and operational expertise to interpret events, weigh advice, assess the developing situation, and modify his plans and orders as required.[3]

Many, if not all, of these same activities must occur in a wargame. What differs, however, is the environment in which they take place. Aside from the fact that actual forces are not maneuvering and shooting at each other, perhaps the largest artificiality of the gaming environment is its generally poor re-creation of communications and data-handling facilities, and their effects on the quantity, quality, and timeliness of the information available to the commander. Even the elaborate Naval Warfare Gaming System uses only a relative handful of communications circuits, status boards, and displays; and data-transfer rates are often deficient when compared to high-speed, real-world systems. On the other hand, the communications that are available in most professional wargames are generally highly reliable and seldom interfered with by the enemy. These problems are exacerbated in most hobby games because the number of players available usually precludes the establishment of any sort of formal command structure. Indeed, most hobby games are designed for two people to play, one on either side. As a result, many of the key interpersonal relationships of command are abstracted away from the players' experience in the game.[4]

In addition, complete re-creation of the actual processes by which commanders receive information in battle appears impractical in a wargame, even in the most advanced professional

252 types. Most obviously, the commander can never go to the front line of his fighting forces or the battle bridge of his ship and see things for himself. It may be feasible, however, to re-create much of what the commander would be able to learn about the situation, even if the details of how he learns it must remain imprecise. For example, a battle-group commander may best learn about the course and status of an antisubmarine warfare operation from information provided by his staff or by monitoring the antisubmarine warfare coordination communication circuit. By keeping track of the information that might actually pass over such a circuit, as well as the errors and delays associated with it, a wargame designer might be able to structure a game system to allow the commander to obtain the appropriate information without actually re-creating the circuit. Such "design-for-effect" techniques can be useful when a detailed re-creation of precise mechanisms is beyond the scope and capabilities of a game. (Such an approach is fairly common in hobby gaming, with games such as *Squad Leader* and others designed by John Hill being the principal examples.)

In a similar manner, complete re-creation of the entire staff and command structure appropriate to a player's decision level is virtually impossible in most wargames. As a result, to define the appropriate roles for various participants, the game's designer must carefully consider the game's objectives, scope, and level of player activity. A summary of the types of tasks important to game play should be given to the players, although many players will prefer to allocate tasks to their assistants as they see fit. Some predefined structure of this sort can help players avoid overlooking critical staff functions whose performance would be impossible to ignore or forget during an actual combat operation. In hobby games, this structuring task is accomplished by the sequence of play; in many professional games it can be accomplished through the device of the "move sheet," by which players report their decisions to the control team.

Although many artificialities limit the ability of a wargame to simulate the decision-making environment of a combat situation realistically, there are many aspects of a commander's operational activities that wargames can reproduce with a surprising amount of fidelity. The intellectual experiences that result from such activities often reflect many of the critical aspects of similar real-world situations. Thus, a wargame can be not only a good **253** vehicle for teaching lessons about the job of a commander to those inexperienced at it, but can also be an opportunity for commanders to practice the intellectual skills they need to do their jobs well. The activities for which a wargame can provide reliable experiences are listed below.

Areas of Activity Adequately Simulated in Wargame Play[5]

Operational

- Force selection and employment
- Integrating platforms to accomplish a task
- Tactical decision making (at appropriate levels)
- Exploiting platform and system capabilities
- Overcoming platform and system limitations
- Rapidly assessing operational and tactical situations

Command

- Delegating authority
- Articulating battle philosophy, directives, and orders
- Establishing information requirements for decision making
- Devising effective ways to display and evaluate information
- Assessing advice
- Crisis leadership

Scenario

- Exploiting geography
- Exploiting environment
- Exploiting international political relations

PREPARING TO PLAY

In a professional wargame, all the efforts of game sponsors, designers, and developers are directed toward producing a game *system*. The game itself takes shape only when the players enter the scene. The success and value of the game revolves around the players and their decision processes, a fact of which too many players of research games seem to be ignorant. To fulfill this important place in the research task of the game, the players must prepare themselves (and in some ways the game-control staff) for the job at hand.

254   The game's sponsor or the agency conducting the game will typically provide prospective players with preliminary information about the game's objectives, scenario, forces, and any other relevant information as soon as that information is available in convenient form. Players should study this information and ask game personnel to provide more details or clarifications about important or confusing areas.

As the date of a game approaches, those players assigned to principal commands are usually asked to give the game control staff an outline of their proposed concept of operations for the game's opening moves. The control staff will use this concept of operations to help test the game system and prepare the system and staff for actual game play.

Players are also well advised to give careful consideration to what types and amounts of assistance they might require to carry out their game role. When possible, the players should inform the game staff about the manning levels they require or expect for subordinates or player staff, their information needs and the preferred modes of receiving that information both before and during game play (for example, a player may want to receive daily intelligence briefs). If the players expect to need particular reference publications, they should verify their availability at the game site and make arrangements to send ahead or bring with them those materials not available.

Finally, it can be very helpful to game-control personnel if players can identify those types of decisions the players intend to reserve to themselves and those they intend to delegate to the controllers. There are sometimes conflicts between what the players expect and what the controllers are used to, conflicts that can disrupt the play of a game if not ironed out ahead of time.

Many professional wargame players and controllers may be surprised to learn that even hobbyists must prepare for play, at least if they intend to play seriously or competitively. The amount of preparation involved in playing a hobby game has varied over time and with the type of game involved, but there does seem to be a more pronounced tendency for the more recent games to require far more preparation than their ancestors did. It takes more than learning the rules to play a hobby game at all, and much more to play it well. Yet even that most basic step of learning the game's rules can require a great deal of time and effort.

The earliest hobby games, such as Wells's *Little Wars*, had fairly short and simple rules. The players could concentrate on figuring out what they wanted to do, confident in their ability to translate their plans into the mechanics needed to carry them out. The earliest board wargames shared these characteristics. Although when compared to children's games like *Monopoly*, the rules of an early wargame like *Afrika Korps* were complex, they were still not very difficult by modern standards. A few basic mechanics of movement and combat were the heart of the game, and once he had mastered them the player could concentrate on *strategy*.

As players and designers became more experienced, however, their increasing desire for greater accuracy or realism in the games translated into a drive for greater detail in game parameters and processes. The result was longer and more arcane rules. Wargames such as *1914* stopped being just an unusual form of adult game and became a specialized field of endeavor for the experienced, "hard-core" wargame buff. This tendency toward detail, complexity, and exclusivity became most pronounced in the genre known as tactical games.

Tactical games deal with the actual battlefield performance, not of large units whose ability to employ the full range of their weapons may be subsumed in a single, simple "combat factor," but with much smaller units, or even individual men or machines. In these games, the players had to coordinate the detailed employment of their forces—which tank would fire on the house, or which type of shot would be loaded into the broadside guns of a ship-of-the-line. Strategy alone was no longer enough. The player now had to learn how to make his forces fight effectively given the limitations of their physical capabilities. It was one thing to "order" a task force to bombard Midway Island; it was quite another to orchestrate the maneuvers and firings of a long line of square-rigged frigates to break the enemy's line of battle while tacking upwind. The former required only the understanding of what needed to be done; the latter also depended on knowing how to do it.

The general rise in the complexity of all games coupled with the increasing popularity of the tactical level began to make the serious hobbyist into an even greater student of history and military art and science than he had been before. As the details of the process became more important to the play of the game, the essence of successful play became the need for experience, not just

256 in the game's rules and systems, but in the underlying elements of reality that the game attempted to represent.

The difficulty for the game designer was to put together a system that could translate the details and processes of reality into game mechanics while preserving the player's focus on reality. It is the same problem that faces the professional community. Representing the processes of reality in great detail requires complex models; forcing the players to deal with complex models requires that they learn not only what works best in reality, but also how to implement the real solutions, in detail, according to the game's representation. This latter pressure tends to drive players, especially players of professional games, to "game the game" when the latter's representation of reality is, as it must be, less than perfect. In turn, this attitude can lead to a loss of that critical suspension of disbelief, a dropping of the players out of their game roles, and ultimately to a serious reduction in the substance of what the game can accomplish.

There is a strong and in many ways healthy tendency on the part of the professional wargame community to focus attention on the insights derived from the play of a game, that is, its substance, and to limit discussion of the game's mechanics, or process. It is important to remember, however, that the game process is integral to the generation of its substance, and that the former can affect the latter in at least three significant ways.

First, the limited amount of time available for playing the game will determine how much of a battle, campaign, or war the game and its players can explore. These time limitations may also distort the flow of decisions and events by giving decision makers too little time to evaluate and select from the many alternative courses of action that may be available to them during the play of the game. Second, the fact that certain systems, tactics, and operations may not be adequately modeled in the game's methodology may constrain the players to make, or not make, important types of decisions. Finally, the parameters chosen for modeling in the game system, and the types, quantity, and quality of information available to the players will largely determine the basis for their decisions, focus their attention, and possibly even alter their evaluations of capabilities and methods of operation in ways not entirely consistent with their experience or with objective reality.

PLAYING THE BAD GUYS

Because so much of the learning derived from playing games is based on the dynamic interaction of the competing ideas and wills of its players, the wargame player needs an opponent (even if it is only himself, an habitual problem with hobby gamers because of the shortage of live opponents). Many of the insights derived from game play thus depend on the abilities of the opponent. Sometimes it can be very difficult to distinguish insights that might apply in general from those that result from the specific circumstances of that specific game with that specific opponent. In particular, how and why the opponent employs his forces as he does is often the most critical element of learning, and also one of the most difficult to interpret.

It used to be that hobbyists simply did not care about such things. After all, the game was supposed to be fun, and the learning that could result from playing it was to be found largely in its information content and the manipulation of the player's own forces. The situation and opposing forces were postulated by history or the game's designer, and how each player used those forces in that situation was entirely up to him.

The professionals could not be quite so cavalier. A central concern in any serious game is the extent to which players "mirror image" the opposition. There is always an underlying uneasiness about the extent to which an American can "play" a Soviet without simply using Soviet forces and capabilities in the style of an American. As a result, the requirements for accurate representation of "the threat" in professional games has become a topic of some interest, generating a series of papers as well as conferences on the subject.[6]

Playing non-U.S. or threat roles in a professional wargame is not really much different from playing friendly or "Blue" roles. Playing the threat well, however, requires special effort, and often special training or expertise. "Red" players must understand not only the technical capabilities of the opposition, but their tactical and strategic doctrine as well. To "play Red," the player must learn to "think Red."

For this reason established gaming facilities usually have a special team of Red players, typically drawn from the intelligence community. At the Naval War College, for example, there is a

258 special detachment of the Naval Operational Intelligence Center whose responsibilities include playing Red in accordance with intelligence projections of enemy capabilities and intentions.

Red players must be careful, however, not to restrict themselves to the standard or accepted responses to every Blue action. Where uncertainties and debates exist, different approaches can be used in different games or even in different portions of the same game. When specific situations seem to call for slightly more imaginative responses, Red players should sometimes be willing to deviate from overly rigid interpretations of enemy "doctrine." When they do so, however, they should carefully explain their rationale and inform the players that what they did may not be in accord with strict intelligence interpretations of likely threat behavior. Of course, any such unusual actions must be consistent with the game's objectives or they are likely to be disallowed by the game's sponsor.

As hobby wargamers have become more involved in modern-era games, there has been a greater emphasis on limiting player options through the use of restrictive "doctrine" that requires players to deploy and employ their forces in particular ways that are supposedly consistent with the actual practices of the military service and nationality they represent. In some ways, although perhaps more formally, the hobby is thus taking an approach to "playing Red" that is quite similar to that of the professionals.

Unfortunately, this tendency may not be a positive trend in either community. There is certainly a need to avoid mirror-imaging in order to avoid misinterpreting both the capabilities and possible intentions of potential adversaries. What is less clear is the efficacy of imposing rigid restrictions on players as the means of achieving this goal. There is a delicate balance between creating a reasonable representation of opposing capabilities and doctrine and imposing an artificial constraint on the imagination and creativity of the players.

The Western view of the Soviet military as one dominated by rigid doctrine has made such approaches as described above acceptable to many professionals as well as hobbyists. The danger of a too literal and unquestioning acceptance of this view and its translation into wargame mechanics or procedures is that it may lead to just as many self-fulfilling delusions as the mirror-imaging approach. Nothing that forces, objectively, bad or even questionable game play should be acceptable, and in fact will not be acceptable to the players who are forced to operate under such restrictions.

Whether it is a hobby gamer or a professional, to whom winning and losing isn't supposed to matter but inevitably and happily does, the player who loses a game because in his eyes the system prevented him from doing what he had to do to win learns nothing from the experience except that the game is a bad one. Similarly, the player who wins the game learns how to win the game, but not necessarily what is important to victory in reality. The essence of the problem remains the need to provide the players accurate information about their situation and forces, and synthetic experience about how to use those forces in that situation. Artificial doctrine that forces players to act in a certain way without making them understand why that behavior is or is not appropriate is counterproductive.

POST-GAME COMMENTARY

Nearly all navy or other professional wargames conclude with a "hot wash-up," during which some or all of the players are asked for an overall assessment of the game and for specific comments about particular issues or aspects of play. Some games, in addition, allow for an ongoing series of such sessions during game play. Players should make the most of these chances to influence directly what the game produces.

Hot wash-up briefings and other commentaries are a player's major opportunity to discuss the key elements of his decisions with the other players and game analysts. They also give the player the chance to address directly the sponsor's objectives for the game. Because of their importance, players should attempt to base their presentations on notes taken during or immediately after the heat of the action. Such a "battle diary" can help players distinguish their actual thoughts and rationales during the game events from post-event analysis and "hindsighting."

Comments about the mechanics or process of the game should also be a part of the player's post-game commentary. If invited for open discussion, such comments may be appropriate for a hot wash-up. In other cases, players may be asked to fill out comment sheets. All such remarks are helpful in the refinement and further development of games and game systems. The most

260 useful comments are usually those that can be related to specific game situations and show the effects the process of play may have had on the substance of game insights.

Hobby games, too, benefit from the reactions and comments of their players. Although many game publishers include a card in their games through which they solicit the responses of the players to a particular game, most game feedback comes from direct, informal contact between a game's designer and its players (such as at a gaming convention). In addition, there are a number of hobby magazines published on a monthly or bimonthly schedule that contain critical reviews of many recent and older wargames or miniatures rules systems.

8

ANALYZING WARGAMES

Earlier chapters have described the principal duality of a war- **261**
game: in its design or structure, it is a tool for the modeling and
exploration of human decision processes in the context of military action; in its play, it is the use of the tool to carry out such
research. Wargame analysis is similarly split into two distinct but
intertwined strands—the analysis of game design and the analysis
of game play.

In the world of professional wargaming, the focus is on the
analysis of game play. The evaluation of a game's design receives
little attention unless it is apparent that the structure of the game
is having a strong influence on the play of the game (recall the
discussion of process and substance in the previous chapter). Unfortunately, it is often the case that the game's design or structure
really is having such effects on the game's substance, but that
those effects go unnoticed.

Professional game analysis generally follows one of two approaches. The first approach, and the one most likely to be fruit-
ful, focuses on why players made certain decisions and why, in
turn, those decisions led to particular sequences of game events.
Such an investigation should examine the important driving
characteristics of the scenario, the rationales for each side's ac-

262 tions, and how alternative choices might have changed the course of events. This focus on the decision-making process is clearly the one most appropriate for analysis of the play of a professional wargame.

In some cases, however, wargame analysts attempt to treat each of a game's events "as a source of scientific evidence on matters of research interest, such as tactics, employment of new platforms or weapons systems, and certain organizational or procedural ideas."[1] This type of investigation treats the outcome of each game event as a single data point arising from a scientific experiment, and collects many such outcomes from many such games into a single body of evidence.

Investigations that focus on decision making employ the techniques of good analytical history and are most important and appropriate in professional wargame analysis. Investigations that focus on outcomes are often quantitatively oriented and more closely resemble scientific inquiry; they are also less applicable to wargame analysis. In some cases, however, analysts may gain valuable insights by evaluating a series of games, but in a qualitative way. This approach is used at the Naval War College to "surface issues" and identify developing trends.

In the gaming hobby, there is quite a body of analysis of game system and design, but there is very little formal analysis of game play in the professional sense. Although each hobbyist may reflect on his experiences at the gaming table to derive some insights into the underlying realities the games attempt to represent, there is little evidence of attempts to push such informal personal analyses of hobby games in the direction of becoming a serious tool for research into historical or contemporary military affairs.

There have been some rare exceptions, in which hobby games or gaming techniques have received serious academic attention. For example, Ms. Helena Rubenstein, former president of the West End Games Company, designed a game dealing with Robert E. Lee's campaign into Maryland and Pennsylvania during the summer of 1863 (culminating in the battle of Gettysburg) as a substitute for a university thesis. A simplified version of the game later appeared in print for sale to hobbyists (*Killer Angels*). In the 1960s, gaming techniques were used in many major universities, but seldom as tools for historical research. The development of hobby-gaming techniques over the past twenty years warrants renewed examination by the academic community, **263** especially now that experienced gamers are members of that community.

Formal analysis of hobby-game play is generally limited to the type of game-replay articles described in a previous chapter. Even in these analyses, however, the emphasis is on improving the player's ability to play the game, not on deriving insights about the reality the game is supposed to represent.

The bulk of hobby-game analysis is found in articles dealing with game criticism or review. Such analyses focus on the effectiveness of the game design or game system as both a model of the real world and as a recreational device. The two principal criteria by which reviewers judge hobby games can thus be described as the two principal elements of wargaming—realism and playability.

Hobby-game reviews provide, first and foremost, a sort of consumer's guide to the numerous games available to the hobbyist. The reviewer's goal is to help the prospective purchaser decide whether the game is a good one, one worth spending money on, or a bad one, which is better left sitting on the game-store shelf than spread over a player's gaming table. To serve this function, game reviews must provide the reader with certain basic information: the topic of the game; the scope, scale, and level of play; the type, number, and quality of physical components (or style of computer display); and a summary of the game's principal mechanics. Because a game is meant to be played, however, and not just read, the reviewer should also impart a sense of the experience of play—the types and scope of decisions, the level of excitement and intellectual challenge, the ease with which the player can perform required tasks, and the amount of time necessary to play the game to a conclusion. Above all, the reviewer must address the question of whether the game works as a game.

But because wargames are not mere abstract diversions, the game reviewer must also comment on its accuracy or "realism." In this case, the goal is to evaluate how well the game re-creates the historical environment of the events or situations it is attempting to portray. Such evaluations are much more difficult because they require the reviewer to have a good understanding of the historical background underlying the game. They are made doubly difficult by the stress on how the game plays, because too many designers, reviewers, and players operate under

264 the false impression that a wargame cannot at once be an accurate "simulation" and a playable "game." Fortunately, an experienced reviewer, even though relatively unfamiliar with a game's historical setting, can still contribute to a player's evaluation of the game's accuracy. By being scrupulously honest about his own qualifications and carefully explaining his questions and opinions, the reviewer can sucessfully point out those areas of the game he believes are particularly successful or particularly unsuccessful at depicting history. Most importantly, the reviewer must make the reader understand exactly why he holds the opinions he does.

In addition to making historical evaluations, experienced game reviewers often find themselves in a position to comment on a particular design's contributions to the "state of the art." The reviewer can discuss how the game addresses the classic problems of player interaction, movement, combat logistics, and command. He can discuss the new or innovative ideas presented in the game and evaluate how well those ideas and systems are integrated into a workable whole. Such "system review" can be the most difficult of the game critic's tasks, and it is made even more so by the lack of a "recognized consensus as to what the range of design approaches are, what classes of variables there are to handle, and which of these are more or less difficult to handle or process than others; whether it is better to centralize or decentralize information; etc."[2] To overcome this lack of consensus, the reviewer must often define his or her own criteria and, again most importantly, explain them and their basis clearly.

Ultimately, evaluating a game fairly and usefully demands that the reviewer assess the designer's success in achieving his goals. To some, "*a reviewer can only determine if the goals set by the designer have been achieved.* That is the entire scope of a game critic's judgments about any design other than his own."[3] Such a viewpoint, though widely held, especially by game designers, is far too constricting on the game critic and depends far too much on the ability of a game designer to state his objectives explicitly.

Critics have the responsibility not only to comment on whether a designer achieved his objectives, but also on whether those objectives were worth achieving. In addition, when the designer's goals are unclear, the critic has the responsibility to attempt to discern them from the way the game was designed. Such a process of divination will always produce mistakes and misun- **265** derstandings, but as long as the reviewer is honest and clearly distinguishes between fact, interpretation, and opinion, he can make a significant contribution, not only to the game-buying public's data base of consumer information, but also to advancing the state of the art of wargame design. This discussion of hobby-game criticism is summarized below in the form of a model outline for a game-review article.

Model Outline of a Hobby Wargame Review

Introduction
- Subject of the game
- Scope, scale, and level of play
- Components
- Designer's focus
- Overall system description

The Game System
- Principal areas of reality represented in the game
- Important abstractions
- Intricacy of the system, and the mechanical ease of play
- Evaluation of the system's success at achieving the designer's goals and representing the real situation
- Contributions to the wargaming state of the art

The Game in Play
- Scenarios
- Player roles
- Types of decisions required
- Effects of the game system's mechanical requirements on the player's decision making
- Evaluation of the player's experience

Overall Evaluation
- Does the game work?
- Is it a good game?
- Who would be most interested in the game?
- Is the game a good value?

266   The lack of a body of widely available criticism of professional game designs and design practices similar to that found in the gaming hobby may be the single most significant reason that the state of the art of professional design is advancing at a rate far slower than that seen in the hobby games. The question of the realism of a game or particular aspects of it is often alluded to during play. Game reports and briefings make frequent reference to the artificialities of the games. Yet there have been precious few attempts to evaluate the problems of even a single game, much less the entire field, in an appropriate and coherent manner. Books such as Brewer and Shubik's *The War Game*, Allen's *War Games*, or Wilson's *The Bomb and the Computer* confuse games with other forms of research and conceal a few valuable insights in a forest of irrelevancies and misinterpretations. Individual game models have been subjected to detailed scrutiny by operations analysts, but the games themselves are seldom carefully evaluated *as games*.

The lack of serious, competent review and criticism of professional wargames is a significant shortcoming, and one the community would do well to contemplate as one potential factor in the cyclical nature of gaming's popularity and utility in defense research. The lack of time and resources to conduct thorough critical evaluations, driven by the shortage of funds for doing more than actually running the games, is an all too unfortunately real source of the problem. Yet, without such evaluations, the validity of the games is always suspect and their real and potential value too easily denigrated. For the value of a wargame as a tool in defense research, as in academic research, is inextricably tied to its validity as a tool for studying decision processes.

DOCUMENTATION AND ANALYSIS OF WARGAMES

Wargame Validity

Even the participants of a wargame often find it difficult to judge the limits of its validity. For those who are not actually present, assessing a wargame's validity is an even more difficult and complex question. The first problem, simply defining validity, is also the most nebulous. As a start, a wargame's validity can be defined as the extent to which its processes and results represent real problems and issues as opposed to artificial ones generated only by the gaming environment.

Given such a definition, assessing the validity of a wargame's results seems to require answering the following questions:

- How are the game's "results" defined by the participants (when available) or by the available game documentation?
- What outcomes from warfare models define or quantify these results, and how are they obtained?
- How and how strongly do "going-in" assumptions drive results and interpretations; in particular, what is the possible influence of scenario and unstated "subliminal" assumptions?
- How and how much does reliance on "accepted" interpretations of enemy reactions drive the perceived principal lessons of the game?
- How do game mechanics, especially action and reaction capabilities, affect the course of the game and its interpretations?
- How do mathematical models and analyses and the value of the parameters they employ affect game play results and insights?
- How does the occurrence of low-probability events drive perceptions of players during the game and conclusions reported about the game?

Clearly, specific games may require other more specific and technical questions to be answered before the validity of their results can be assessed. If nonparticipants in a wargame are to have a fighting chance of fairly interpreting what that game has to say, however, they must be able to find answers to the questions listed above. The answers to such questions must be available in game documentation, and so these questions must also guide game analysis and documentation from the start.

Wargame Documentation

Just as wargame analysis is closely related to analytical history, a wargame report should more closely resemble an historical treatise than the documentation of a campaign analysis. Just as good historical analysis treats events as indicators of deeper underlying realities, good wargame analysis and documentation treat game events only as indicators of the decision processes of the players. Although game reports describe the major events of game play,

268 they should focus on the underlying reasons for the players' decisions that gave rise to those events. They should also evaluate the extent to which those reasons were driven by realistic concerns or effects rather than by the artificialities of the gaming environment.

The structure of a wargame report should reflect this relative importance of events and causal factors. In general, a simple, straightforward structure is best. It should begin with a short executive summary that outlines the objectives and structure of the game and highlights the key events and insights. Perhaps most importantly, the summary should point out areas or issues raised in the game that require further, more detailed, research.

The report should end with a brief appendix discussing the important elements of the primary models used to support the game. The discussion should explain the roles and relative importance of the various models, describe inputs required, and identify those that drive the results. Where possible, umpire variations of model inputs or outputs should also be discussed, at least for major engagements or classes of engagements. Finally, if models are documented, the appendix should provide references to allow those interested readers a chance to explore the models in more detail.

As shown below, the main body of the report should be built around four sections: introduction or background, game play, insights or issues raised, and conclusions. The introduction should relate the origins of the game to its objectives and its structure or design. It should allow the reader to understand why the game was played and how its results were expected to provide information about specific questions or objectives. The summary of game play should describe the scenario and its effect on play; specify the number of actual players and their roles and commands; and describe in broad terms the courses of events, focusing on player decisions and their underlying reasoning. At times, assessments of the validity of these reasons and the sources of such assessments may also be warranted. The sections on insights derived or issues raised from game play should concentrate on specific matters either pointed out during the course of play or raised in discussion or "hot wash-ups." Where possible, such issues should be keyed to specific game events or classes of events that gave rise to or illustrate the importance of the issue.

Model Outline of a Professional Wargame Report

Executive Summary

Introduction
- Origins of game
- Game objectives
- Game design to meet the objectives

Game Play
- Scenario
- Player roles
- Key events and decisions, integrating the rationales for each

Insights or issues
- Driving factors
- Specific ideas, preferably keyed to major game decisions or events

Conclusions
- Broader insights into major underlying factors
- Topics for further research

Appendix
- Model roles and importance
- Inputs, outputs, and umpire modifications
- Sources of documentation

The first three sections of the report largely represent the historical chronology and causal analysis. The concluding section should identify the deeper factors that may underlie and relate several, possibly diverse, individual insights and issues. Sometimes such themes are not to be found; if discernible, however, such broad insights can prove to be the major contribution of a game to the accomplishment of the objective. Finally, the conclusions should also identify those issues raised by the game that are both important enough and tractable enough to be addressed constructively by further research. In this way, the game can help direct the attention of other defense analysts toward high-priority topics.

270   Perhaps the biggest problem in wargame documentation is the need to balance the speed with which a game report is produced and the amount of time necessary to reflect on and assimilate the insights the game may provide. The pressure to produce "results" quickly is similar to that experienced in exercise analysis, and similar care must be used in game analysis.[4]

In many cases, the purposes of the game will dictate the relative importance of speed and depth of analysis. Training games clearly require almost immediate feedback if they are to be most useful. Games conducted more for the purpose of research may have the luxury of several months of careful analysis before a report is required. Most games, however, are probably well served by the same device seen in exercise analysis: a "quick-look" report touching on the highlights and produced quickly, followed by a more thoughtful and careful full-scale report.

Both quick-look and full reports should follow the outline proposed above. Most of a game's primary objectives should be addressed in the quick-look report. The follow-on report can cover those issues in more detail, raise new ones in unexpected or tangential areas, and address other topics resulting from additional reflection and analysis.

Sponsors of games played at the Naval War College can generally count on the War Gaming Department to provide some support for the production of a quick-look report. However, because the department is not staffed to provide extensive analysis for every game, sponsors must supply their own analysts if they want more detailed follow-on reports.

Wargame Analysis

Plans for game analysis must be made as early as possible in the gaming cycle with a view toward producing a report structured as outlined above. Ideally, the game sponsor, designer, and those responsible for game analysis should jointly determine the objectives of the game, the overall shape of the design that will allow those objectives to be met, and the data and information that must be collected to meet those objectives. Early joint discussions of objectives, mechanics, and analysis can help prevent the potential problems of designing a game that addresses the wrong issues and structuring a game analysis that focuses on the wrong measures.

Typically, those game participants responsible for analysis **271** are the game's only historians. During the course of play, analysts must record the major decisions made by the participants, their rationales, and the game events to which the decisions led. The analysts are also responsible for cataloging the key insights that participants derive from game play and discussion.

The functions of observer and data collector are only a part of the analyst's responsibilities, however; in essence, observations and data collection provide only the raw material for analysis and synthesis. The analyst must go beyond merely what happened in the game to understand not only the immediate causes of events, but also the deeper themes that may underlie an entire chain of cause and effect. This analytical process differs from the quantitative "scientific" analysis to which that term is most often applied. The raw data of a wargame is the interaction of players and their decisions; it is history, not science. The process of game analysis is much more akin to exploratory research and historical analysis than to the evaluation of physical experiments or systems analysis. On the other hand, most wargames use the conceptual, if not always the mathematical, models as often found in operations research or systems analysis approaches to defense problems. Thus, the analysts must be well versed in the characteristics, capabilities, and limitations of such models to understand how well or poorly they represent the reality of military operations and how appropriately they affect the play of the game.

To analyze the game effectively, the analyst must keep in mind the entire gaming process, from the initial formulation of the game's objectives, through design, play, and documentation, and even to the interpretation of results by nonparticipants. Wargame analysis is a complicated process requiring careful observation, questioning, thought, and synthesis of events and insights; it is essentially the art of discerning order in the midst of chaos. To do a credible job of game analysis, the analyst must have a coherent idea of what to look for and why before being thrust into the game itself. The appendix lists several questions that a wargame analyst may wish to consider in his preparation for and participation in a game. These questions are only a guide; there is no magic formula for doing good game analysis. Each game is different, and what is important in one may be unimportant in the

272 next. Hopefully, however, the principles discussed in this chapter and the questions in the appendix will help the new analyst sort out which is which. Only good analysis and documentation can allow valid interpretation of the game and application of its lessons to real-world problems.

9

INTEGRATING WARGAMES WITH OPERATIONS ANALYSIS AND EXERCISES

Wargaming is one of the U.S. Navy's principal tools for educating **273** its people and for evaluating its combat capabilities. This latter process is crucial today when so many weapons, systems, and ideas are untested in combat. Short of actual military operations, the navy makes use not only of wargames but also exercises and operations or systems analysis to help it learn about the strengths and limitations of its strategies, concepts of operations, and tactical and technical capabilities. Too often the layman will equate all three of these tools (indeed, often referring to all of them by the generic term *war gaming*). Just as often, it seems, the professional defense specialist acts as if the roles of wargames, analysis, and exercises overlap to a great extent; frequently such specialists seem to view those tools as functioning independently of one another or even in competition with one another.

Such a shallow view not only fails to distinguish the subtle but important differences among wargames, analysis, and exercises, but also impedes their correct use. Only by integrating the information that can be extracted from the proper application of all three processes will the navy and other elements of the defense

274 community obtain a balanced and well-rounded understanding of the potential problems and opportunities of future combat. In order to perform such integration, it is first necessary to understand how wargames differ from exercises and analysis, and to explore the inter-relationships and complementary nature of the three processes.

DEFINITIONS

Earlier we defined a wargame as a warfare model or simulation, not involving actual military forces, and in which the flow of events is affected by and in turn affects decisions made during the course of those events by players representing the opposing sides. The key words in this definition are players and decisions. Wargaming is an experiment in human interaction. Without human players there may be a model, but there is no game. A true wargame is best used to investigate the decision processes of its players and how those processes interact; it is not well suited to the calculation of outcomes of physical events.

Analysis or "operations research," on the other hand, has been defined by two of its founding fathers as "a scientific method of providing [decision makers] with a quantitative basis for decisions."[1] Here, the key words are scientific and quantitative. Because the field of analysis has grown so large and diverse (including under its aegis systems analysis, operations analysis, and even at times engineering and policy assessment), many other definitions of it have been proposed. In a textbook written for use at the U.S. Naval Academy this "more thorough modern definition" is proposed: "Operations analysis is the application of scientific knowledge toward the solution of problems which occur in operational activities (in their real environment). Its special technique is to invent a strategy of control by measuring, comparing, and predicting possible behavior through a scientific model of a situation or activity."[2]

As discussed in chapter 3, modern defense analysis was born during World War II when both the British and Americans made concerted efforts to mobilize the scientific community in support of military operations, not merely in the development of new technologies. The efforts of these early pioneers, such as Morse and Kimball in the U.S. and P. M. S. Blackett in the U.K., were directed at problems of antiaircraft operations, convoy protection, and antisubmarine warfare.[3] Wartime groups such as the British Operational Research Sections and American Antisubmarine Warfare Operations Research Group evolved after the war into formal analytic organizations. Analysis received its biggest boost in the U.S. under Secretary of Defense Robert McNamara in the early 1960s. Although the influence of analysis has had its ups and downs since those early days, it remains a major weapon brought to bear on both sides of many defense debates. Despite the divergence of views on the nature and proper roles of analysis, however, virtually everyone in the defense community agrees that the fundamental characteristic of the discipline is its scientific and quantitative nature.

Military exercises are fundamentally different from both analysis and wargames. Exercises can be considered any activity involving the operation of actual military forces in a simulated hostile environment. Here, the key words are forces and simulated. The navy conducts many different kinds of exercises to achieve different objectives, but all true exercises are characterized by real-time operation of ships and aircraft, which often expend real or simulated weapons against some "enemy" force. Oftentimes, this "enemy" is represented by U.S. or allied forces tasked to play the role of some hostile power.

It is clear from the above definitions that wargames, exercises, and analysis are distinct approaches to studying military operations. Although often related and in some ways similar, each of the three tools focuses primarily on different aspects of warfighting reality. As a result, each technique can be an effective device for learning about specific aspects of a problem, but none gives a completely balanced view of reality.

The physical sciences are the paradigm of analysis. Analysis focuses on the physical processes of reality, adopting a philosophy of approximating those processes with mathematics that can, at least in some sense, be "solved." Analysts build mathematical models of some major elements of reality, make measurements to quantify the parameters of the models, and then manipulate both the models and the parameters to learn about reality or to find the best solutions to the problems it poses. Although the mathematics used in analysis is objective, the choices of which models to use and which parameters are most important, the assumptions that underlie the analysis, and sometimes even the method of

276 solving the mathematical problem can all be subjective. As a result, it can be very difficult to translate valid learning about the model into valid learning about reality. In making the translation, analysis must simplify and often discard much that is not reproducible or readily predictable—including at times, human behavior.

Wargames, on the other hand, focus precisely on human behavior, particularly on human decision making. The learning that comes from wargames comes both from the experience of making decisions (playing) and from the process of understanding why those decisions are made (game analysis). The outcomes of game decisions are frequently defined by the results of mathematical models that are often similar or identical to those used in analysis. In a wargame, however, these models are used in a fundamentally different way. The models used in a wargame are typically stochastic in nature—that is, they may produce a variety of outcomes, and the specific one that occurs for any use of the model is determined by some chance device that reflects the probability distribution of the various outcomes. Thus, the "roll of the dice" provides a wide range of possible results or snapshots of reality with which the players must deal. It is in this sense that model results should be considered inputs to wargames, whereas such results are more often considered the outputs of analysis. Wargames do not do very well at producing quantitative measures because they are often little more than a single realization of a complex stochastic process. Instead, the value of a wargame lies in qualitative assessments of why decisions are made. Thus, to exploit the power of wargaming, the physical sciences must give way to a new paradigm, that of history. People and decisions become paramount.

Exercises focus on doing. They are primarily tools for training and are usually designed with training goals uppermost in priority. Operational and tactical decisions are sometimes constrained because of requirements to exercise systems and train personnel. Even unscripted free-play exercises are usually restricted because of safety requirements or geographic limits on the exercise operations area. Exercises are often viewed as experiments that can provide reliable data about how things "really are" so that the models used in analyses or games can be calibrated or made more realistic. In many cases, such a view is a useful one, but it is one that requires care in the process of interpreting num- **277** bers whose origins and import are sometimes difficult to judge. There is no generally accepted way to accurately adjust the results of an exercise to reflect its artificialities. Thus, in order to focus on the execution of orders and tasks, exercises must often restrict the range of physical parameters and processes that can be explored and limit the potential decisions the participants can make. As with analysis and wargaming, the actual results or outcomes of the execution of exercise activities (particularly combat activities) can only be approximated. Exercises, too, are not real.

WARGAMES AND EXERCISES

As indicated in earlier discussion, perhaps the easiest way to distinguish between a wargame and an exercise is that the latter involves the actual movement and operation of military forces. An exception to the rule is the command-post exercise or CPX. The individual services, the joint chiefs of staff, and even the civilian elements of the National Command Authority conduct command-post exercises to test procedures and identify potential problem areas that could arise in an actual crisis. Major command-post exercises like Nifty Nugget may see some movement of small units or even individual reservists recalled in a test of the mobilization plans, but most CPXs do not involve ships putting to sea or army battalions deploying to wartime positions.[4] In fact, the careful scripting of the major events in such exercises makes them very similar to the types of one-player wargames described in Chapter 4.

A true exercise requires military units to operate as if they were in a real-world situation, with obvious differences. The principal focus of an exercise, as described earlier, is on training the participants to function more coherently as a team. An aircraft carrier battle group that is about to leave U.S. home waters to deploy overseas will usually go through a battle-group evaluation exercise, to practice coordinated operations and identify areas that may require additional training before deployment. Although each exercise is an opportunity to conduct operational research, and military and civilian observers and analysts usually record and reconstruct the major exercise events to identify important "lessons learned" or new operational insights, for the most part research efforts must take a back seat to training.

278   Wargaming has also played a major role in training and education. Indeed, much of the early emphasis on wargaming at the Naval War College centered around its ability to provide players a taste of the intricacies involved in maneuvering large battle fleets, an opportunity that few would have in peacetime. Yet even between the world wars, and especially since 1945, wargaming has become more and more popular as a tool for exploring a broad range of strategic, operational, and tactical questions, especially those focusing on the decision-making process.

In addition to this different balance of training and research emphasis, there are other differences between wargaming and exercises. The major differences can be summarized in terms of cost, time scale, flexibility, level of play, nature of the participants, and characterization of results.

Compared to the cost of running an exercise, a wargame is usually quite inexpensive. The actual play of the game seldom involves more than a few dozen officers and some number of supporting technicians, clerical workers, umpires, and analysts, and only for a few days. Although pregame planning and postgame analysis efforts may require several months of work, they involve only a relative handful of people. A major exercise, on the other hand, usually involves thousands of military and civilian personnel, from the technician who mans the radar to the public relations officer who explains the goals of the exercise to the press. It also requires the operation, logistical support, and maintenance of large numbers of ships, aircraft, and other equipment for periods of up to several weeks. As a result, the costs of a wargame and an exercise that deal with the same general topic can differ by several orders of magnitude.

A somewhat dated but still valid example of such cost differentials is given by Hausrath. Based on various government estimates, Hausrath concluded that the United States had spent about $246 million per day during World War II, and about $16 million a day in Korea. During the Vietnam conflict, the cost had ranged from $5.5 million a day in 1965 to about $70 million a day in 1968. "The daily cost of operating a game that would be representative of conflict of the type described above amounts to about $500 to $4,000."[5]

As for maneuvers and exercises, considering only additional costs incurred above normal operating expenditure, Hausrath listed costs for four exercises. These costs ranged from just over **279** $541,000 for a two-week command-post exercise (nearly $40,000 per day for a full fourteen days) to a high of some $60 million for a massive army–air force exercise in 1964 called Joint Exercise Desert Strike (roughly $4 million a day).

Because a wargame does not employ actual military units, the control group can regulate the advance of time during the play of the game to run much faster or much slower than real time. A game such as the Naval War College's Global War Game, which explores the strategic options in a long, worldwide war, may have the game time advance at a rate ten times that of real time. (That is, one day of game play represents the activities of ten days of war.) Alternatively, a game such as NAVTAG, which explores the tactical options available to the commanders of individual ships or battle groups during a few hours of combat, may slow time down so that one hour of game play represents only a few minutes of real time, thus providing the players more of an opportunity to analyze and appreciate a complex tactical situation. Exercises, for the most part, must be carried out in real time. Some "time jumps" between phases of an exercise are possible, but the actual activities of the units involved can seldom be done at anything other than real-time rates.

Because of the administrative, logistical, and sometimes even political complexities of staging large exercises, they are typically carried out at the lower levels of battle groups or individual platforms (or army divisions or battalions). Theater or major operational level exercises like the navy's FLEETEX (involving multiple battle groups, submarines, and land-based aircraft) or NATO REFORGER (involving many divisions from all of NATO, including air and sea transport of reinforcements to the Continent) are held only infrequently. Wargames, on the other hand, can be played easily at any level of force or violence—up to and including that of the National Command Authority and global strategy and policy in a worldwide NATO/Warsaw Pact war.

Unfortunately, because senior civilian and military leaders can seldom absent themselves from their real-life responsibilities for any prolonged period, active participation in exercises is usually restricted to lower-ranking military officers, seldom including fleet or theater commanders. Political background and diplomatic decisions are often simplified or assumed away in such

280 exercises. In many wargames, on the other hand, civilian players are involved. Their representation of political authorities (as opposed to military men acting as politicians) can often add new and quite different perspectives to those of the military participants, with sometimes surprising and at times frustrating results. (For example, an elaborate military plan to operate fleet units in the territorial waters of a NATO ally prior to the opening of hostilities might be frustrated by the political decision of a civilian "president" who finds the military reasons for adopting such a plan insufficient to counterbalance the potential diplomatic problems its implementation could entail.) Unfortunately, the same difficulties that prevent most exercises from enjoying the prolonged involvement of senior decision makers plague wargames as well.

Finally, as discussed in detail earlier, the results of wargames are best characterized as qualitative assessments of why and how decisions were made. Exercise results are more often characterized as quantitative summaries of unit performance. Wargame analysis documents decisions. Exercise analysis measures operational parameters such as system availability, speed at which orders are transmitted and executed, the number of detections made as a fraction of opportunities, the percentage of targets engaged, or others. Table 4 summarizes the comparison of exercises and wargames.

WARGAMES AND ANALYSIS

On the surface, wargaming appears to have even more in common with systems or operations analysis than it does with exercises. Both wargames and analysis are, in some sense, an intellectual abstraction of warfare without the bother of maneuvering real people and machines over real terrain or oceans. Both wargames and analysis use scenarios to set the ground rules for and structure the research. Data bases provide both wargames and analyses the basic information they need about the physical parameters and processes of the situations likely to be of interest. To simulate how these parameters and processes interact in the real world, both wargames and analyses employ mathematical models and some set of rules and procedures that assure the logical flow of cause and effect. Despite this surface similarity, however, in both their goals and their operation, wargames and analyses differ significantly.

Table 4. Comparison of Exercises and Wargames   281

	Exercises	Wargames
Activity	Operation of actual forces	Simulation of operations
Goals	Training; evaluating performance	Training; exploring decision processes
Cost	Expensive	Relatively inexpensive
Time scale	Real time	Adjustable
Flexibility	Resource-constrained; limited by availability of forces	Require relatively few resources; may be played nearly anytime or anywhere
Levels of play	Primarily tactical with limited operational	Tactical, operational, strategic—all possible
Participants	Military; seldom highest ranks	Both military and civilian; seldom highest ranks
Results	Quantitative measures of performance	Qualitative assessments of decisions

In the defense community, the term *analysis* usually connotes systems or operations analysis. As described earlier, such analysis may be characterized as a collection of techniques for quantifying and manipulating quantitative information about physical parameters to calculate the quantitative outcome of physical processes. Wargaming, on the other hand, is a tool for exploring the effects of human interpretation of information rather than those of the actual information (or data) itself. Wargames focus on the decisions players make, how and why they make them, and what effects they have on subsequent events and decisions.

In the diverse world of analysis, the type that most closely resembles the scope and form of wargaming is what may be called classical campaign analysis. Thus, a comparison of these two techniques best highlights the differences between wargaming and analysis.

A campaign analysis, not surprisingly, analyzes a campaign, defined as a sequence of tactical encounters between opposing forces, with the results of each encounter determining the potential numbers and types of forces available for the next. One of the most frequently analyzed campaigns in the 1970s was that for

282 control of the Atlantic sea lines of communication during a NATO/Warsaw Pact war. In such a campaign, Soviet submarines and long-range bombers were opposed by NATO land-based fighters and antisubmarine warfare aircraft, aircraft carrier battle groups, nuclear and diesel submarines, and convoy escort groups of surface ships and helicopters. Various Soviet strategies for deploying their forces were opposed by different combinations of allied antisubmarine barriers, sweeping operations, and convoying policies. Complex mathematical models represented everything from the ability of a sonar system to detect a submarine to the results of a coordinated attack of several missile- and torpedo-firing submarines on an Allied convoy protected by fixed-wing aircraft, submarines, helicopters, and surface ships in a layered defense in depth. A large bookkeeping model kept track of the types and numbers of forces lost on both sides over the course of a ninety-day campaign. The army and air force employ similar campaign analysis techniques to examine the course of a war on NATO's Central Front, or the balance of power in the air.

When carefully structured and thoroughly carried out, such campaign analyses might be expected to yield valid insights about the following:

- The feasibility of alternative and opposing strategies
- Areas of strength and weakness in combat power and sustainability for both sides
- Factors and parameters that critically affect the results of the campaign and the sensitivity of the results to such factors
- How different types and numbers of forces can be used to best advantage
- The relative contribution of the various types of forces to the overall outcome of the campaign

The process of carrying out such a campaign analysis usually requires the analysts to define a sequence of events—often simply a string of sequential engagements—and to calculate the expected (or average) outcome of those events based on the postulated mathematical models and information about forces and capabilities. The overall campaign outcome is a set of numbers representing the expected losses to each type of force over the course of the campaign. In rare cases and for special types of forces like aircraft carriers or AWACS aircraft, a probability distribution for the numbers lost may replace or elaborate on the single mean number.

The problem is that the assumptions that led the analysts to **283** define the initial allocation of forces and sequence of events may have been faulty, and so may result in a set of outcomes that would be patently absurd in real life. For example, if all the Soviet submarines that attempted to attack aircraft carrier battle groups were sunk without damaging a carrier, the analysis might be restructured to have fewer submarines allocated to such suicide missions, or to prohibit such attacks after some initial period of war (during which, presumably, the Soviets would learn of the futility of such efforts). Through trial and error, the analysts go back through the event sequences to determine what changes in strategy or tactics could result in a more balanced outcome. The old sequence is then discarded and replaced by a new one. This iterative procedure goes on until the analysts are satisfied that both sides are employing nearly optimized strategies. This final sequence is then used to run the campaign to an analytical conclusion. The final results, usually defined in terms of the expected attrition suffered by both sides, becomes the basis for assessing feasibility or identifying critical factors, and for comparing variations of the assumptions underlying the analysis. For example, the "base case" analysis may assume that all allied military shipping runs in slow but heavily defended convoys, while a variation may allow the ships to sail independently along "protected lanes." Results are then compared to see which strategy had more ships sunk, or which delivered more material earlier.

Wargames allow for the continual adjustments of strategies and tactics by both sides in response to the developing situation and outcomes of specific engagements; such adjustments are not seen in campaign analysis. Wargames afford their players a measure of control over events through the decisions they make during play. Unlike a campaign analysis in which changes in strategy occur as a result of calculating the outcomes of implementing the strategy, wargame decisions are not based on a clear and complete understanding of all the facts (much less the results) but rather on how the players view the facts through a cloudy and possibly incomplete frame of reference that is often distorted by preconceived notions, poor information, and the pressure of time—in other words, the fog of war. In a campaign analysis, a strategy that leads to disastrous losses is simply discarded; in a wargame, most decisions cannot be recalled after they have been made. (In some cases, however, a player may make a particularly

284   bad decision that results in a situation that would prevent further play from achieving the objectives of the game. In such cases the control staff will usually turn the game clock back to the time before that decision was made and allow the player to try again, not, however, before the point has been made about why that decision was such a poor one.)

Although immediate outcomes of wargame decisions are sometimes defined by mathematical models (such as the result of an air attack on a carrier battle group), the true effects of those decisions ripple through all subsequent game decisions and events. What and how much is lost in a particular wargame engagement or campaign is far less important for interpreting the insights the game can provide than how and why those engagements occurred as they did.

The end product of a classical campaign analysis can look deceptively like the outcome of a single wargame. But it is a game in which all decisions are premade, poor decisions are self-correcting, uncertainty eliminated, and chance averaged away. Such campaign analysis (and other forms of analysis) can provide important insights into the effects that systems and tactics might have on the outcomes of combat operations in the particular circumstances assumed by the analysis. Yet, it has enormous difficulty in capturing the dynamic elements of warfare or in illuminating facets of reality not already incorporated into its assumptions and mathematical models. Because analysis tends to focus on the quantifiable and reproducible, on the mean rather than the outlier, it can provide little insight into why and how a brilliant hunch or incredible blunder, a bold gamble or paralyzing indecision can turn carefully crafted plans into beautifully executed fiascoes, or ad hoc operations into decisive victories. There are no Chancellorsvilles in campaign analysis.

While analysis focuses on systems, the true value of wargaming lies in its unique ability to illuminate the effect of the human factor in warfare. By their very nature, wargames seek to explore precisely those messy, "unquantifiable" questions that analysis must ignore. Wargames teach us what we didn't know we didn't know.

To accomplish that, however, wargames must give up any vain hope of achieving the detailed mathematical structure and rigorous calculation characteristic of analysis. A wargame is not and will not ever be a mathematical experiment whose initial **285** conditions can be re-created precisely and varied at will. The fundamental initial conditions of a game—the knowledge, talent, character, and experience of the players—changes as players change or as they play the game more. Unlike analysis, such parameters are not physical and not mathematical and may not be varied readily over a wide spectrum of possible values.

There is another important difference between analysis and wargames. Because analysis requires highly technical and quantitative training, most of its practitioners are civilians, despite the recent increase in the number of military officers who have earned advanced scientific degrees. Although the best of these civilian analysts work closely with their military clients to keep their analyses militarily sound, it is rare indeed to find an analysis in which all major decisions about force employment, missions, and operating concepts are made by active-duty military personnel. Except for that minority of wargames used by civilian analysts for strictly exploratory purposes, most military wargames cast military officers in the military decision-making roles. The differences in perspective and experience can sometimes result in significant differences between how a civilian might address a military problem and how the same problem is handled by someone in uniform. As mentioned in the earlier discussion of exercises, it can also be misleading to have military officers play civilian roles, for all the same reasons.

Table 5 summarizes the comparison of wargames and campaign analysis.

SYNTHESIS

This comparison of wargames to exercises and analysis highlights some of the major similarities and differences among the three principal techniques of learning about defense issues. It also demonstrates that placing complete reliance on any one or two of these techniques can result in a skewed impression of the critical features of the reality of warfare. To gain a balanced view of that reality, it is necessary to integrate the insights available from wargames, exercises, and analysis along with the experience of history and current operations.

The first thing to remember is that wargames, exercises, and analysis involve no actual fighting. Because this is the case, none

286 Table 5. Comparison of Campaign Analysis and Wargames

	Campaign analysis	Wargames
Objectives	Quantitative insights into feasibility, critical physical factors	Training; exploring decision processes
Event sequence	Preordained	Dynamic
Engagement outcomes	Typically expected value	Usually stochastic
Learning	Iterate till balanced outcomes	Few second chances
Interpret	Results	Processes
Participants	Primarily civilians with military advice	Primarily military in military roles

of these tools can capture all of the human elements of real combat. Military history is full of examples in which courage, fear, morale, and leadership provided the decisive determinants of defeat or victory. Wargames and exercises, by requiring participants to process information and make decisions in the presence of uncertainty and under the pressure of time, provide greater opportunities for exploring some of these factors than does analysis, but even their ability to re-create the stresses of combat is limited. The daily hours of play for a professional wargame seldom exceed those of a normal working day, and the players know that at the end of the day they will be back in the real world, and at the end of the week or the month they will be back at their normal duty stations. Even exercises, in which the physical conditions of operation and the physical activities of the participants are more similar to those of actual warfare, can only produce a fraction of the real pressures that affect people's performance when actual weapons may be fired in anger.

Similarly, the effects of such weapons can only be simulated during peacetime. Even live firings of things like surface-to-air missiles take place under controlled conditions that cannot replicate the actual environment of combat. As a result, wargames, exercises, and analysis must all rely on mathematical models, which can only partially account for the effects of weapons on the men and machines of war. The outcomes of engagements, from a submarine-versus-submarine duel to a large-scale air attack

on massed battle forces, are assessed by using th
and interpreting or modifying their results on the basi.
(or analytical) judgment. Unfortunately, because ma
weapons have seen limited or no use in actual con
els and judgments are seldom based on a substantiaɪ body of
hard data.

Finally, there is a tendency, most pronounced in the mathematical world of analysis but extending to a lesser degree even to
exercises and wargames, to seek the ultimate truths of combat
in "typical," "expected," or "likely" results. Yet, if history teaches
us anything, it should remind us that in war the unexpected is
commonplace. Too often the highly detailed engineering or
expected-value models that are touted by their creators and users
as reflecting every important factor in combat actually obscure
what Samuel Eliot Morison called "the tremendous influence of
luck in all warfare, especially naval warfare."[6]

Although there are many other artificialities and shortcomings of wargames, exercises, and analysis, our goal is not to
catalogue all of them. Practicioners of each of the techniques are
(or should be) well aware of their individual artificialities and
strive to improve their ability to overcome or minimize them.
What is too often missing in such efforts is the synthesis of all of
the techniques.

Alone, wargames, exercises, and analysis are useful but limited tools for exploring specific elements of warfare. Woven together in a continuous cycle of research, wargames, exercises,
and analysis each contribute what they do best to the complex
and evolving task of understanding reality. Contrast this philosophy of integration, illustrated by "the cycle of research" at the top
of figure 8, with a philosophy of separation symbolized by the
more traditional "spectrum of analysis," illustrated at the bottom
of the figure.

Rather than using each of the tools in isolation, a great deal
more can be accomplished by employing wargames, exercises,
and analysis to address the individual pieces of a problem for
which they are best suited, and then integrating and interpreting
their separate results into a combined picture. The process for
doing so is difficult and requires special combinations of experience, insight, and training. There is no magic formula, but an
example may illustrate some of the possibilities.

A question of great interest to the navy centers on whether

The Cycle of Research

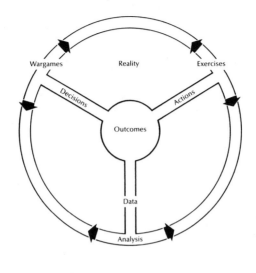

The Spectrum of Analysis

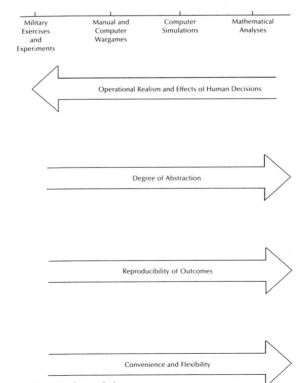

Military Exercises and Experiments	Manual and Computer Wargames	Computer Simulations	Mathematical Analyses

Operational Realism and Effects of Human Decisions

Degree of Abstraction

Reproducibility of Outcomes

Convenience and Flexibility

Figure 8. The cycle and the spectrum.

aircraft carrier battle groups can operate usefully and effectively **289** in specific geographic areas when opposed by a particular type of Soviet submarine threat. Analysis can construct models and devise methodologies to describe the effectiveness of ASW barriers, direct carrier battle group defenses, and submarine attack capability. These models would be mathematical functions of sensor and weapon performance based on the best available theoretical and experimental data. Measures of effectiveness, such as the probability of an attacking submarine's being killed before firing at a carrier, can be defined and calculated on the basis of the assumed parameter values, and the effects of changes in those values can be quantified through the changes in the measures of effectiveness. In this way, the analysis might identify critical physical parameters.

Informed by the results of the analysis and possibly using models adapted from it, the navy could conduct a wargame to explore the concept further. The game could include not only military commanders who might have to execute the operation, but civilian decision makers as well, thereby yielding different points of view and value judgments. Such a game could shed new light on the political ramifications of deploying or not deploying carrier battle groups to the region, the availability of specific force levels under a variety of conditions, the rules of engagement under which those forces might have to operate and how those rules might change over time, and the possibly unexpected reactions of an enemy whose perceptions differ from our own. Similarly, the dynamic environment of a game may cause players to react differently than assumed by a static analysis.

However imperfectly large-scale political and operational decisions are modeled in a wargame, they can sometimes have more important effects on the conduct and utility of an operation than the detection range of a sonar or the probability of accurate weapons placement given detection. Yet, without the understanding of the latter factors provided by good analysis, the decisions can be too abstract, too sterile, and their effects assumed rather than assessed. The gaming and analysis pieces must fit together.

Exercises conducted in the area of interest by forces similar to those envisioned in the proposed concept of operations can often help illuminate how those pieces can or should fit together and also supply some missing pieces of the puzzle. Careful obser-

290 vation, reconstruction, analysis, and interpretation of exercise events and system and unit performance can provide the insights and data to improve the form of mathematical models and the quality of parameter estimates. In addition, the physical execution of maneuvers and procedures required to carry out the operation can help to identify important operational opportunities or potential problems that the analysis and wargaming may have downplayed or failed to consider at all.

In such an integrated approach, each of the tools strengthens and supports the others. Analysis provides some of the basic understanding, quantification, and mathematical modeling of physical reality that is required to assemble a wargame. The game presents some of the data and conclusions of the analysis to its participants and allows them to explore the implications that human decision making may have for that analysis. It can illuminate political or other non-military, non-analytical assumptions and points of view, raise new questions, and suggest modifications to existing or proposed operational concepts. An exercise can allow the navy to test those concepts at sea with real ships, aircraft, and people, in order to measure the range of values that mathematical parameters may actually take on, to verify or contradict key analytical assumptions, and to suggest even more topics for gaming, analysis, and follow-on exercises, thus continuing the cycle of research and learning. It is only by integrating the techniques, not by isolating them from each other, that the navy and the rest of the defense community can hope to gain a better and balanced understanding of the potential reality of future warfare.

PART III

PROSPECTS FOR THE FUTURE

10

NAVY WARGAMING TODAY

SCENE 1

The office of a navy research organization. Two civilian analysts study a map of the Mediterranean, which is overlaid with a grid of hexagons and cluttered with stacks of small colored cardboard squares. Each of the squares is decorated with several numbers and the silhouette of a ship or aircraft. One of the analysts is carefully calculating the chances that a squadron of F/A-18 strike aircraft will be able to penetrate the air defenses of a *Kirov* battle group; the other searches the map for another Victor SSN to deploy to the Straits of Messina.

SCENE 2

The wardroom of a navy destroyer at sea. Four officers are seated around a large table on which rest three microcomputers. Two of the officers are typing commands into their computer, maneuvering their DDGs to detect, target, and attack their opponent before he does the same to them. One of the other officers is watching the screen on the third computer, noting the orders each of the players is giving and contemplating what he will say about the quality of those orders when he makes his postgame assessment.

294 The fourth officer, following a long and respected tradition, is giving advice to all the others.

SCENE 3

A large conference room at fleet headquarters. Fifty officers, from flag ranks to ensigns, fill the room, about a third of them seated at a long table. There are several charts on the walls and on smaller tables, but little other equipment is visible. What is apparent is the intensity of the discussion as the logistical difficulties of supporting a marine landing are explained and potential solutions to the problems are debated.

SCENE 4

Sims Hall, the Naval War College. The building is bursting as some 500 military and civilian participants of the annual Global War Game wrestle with the strategic, operational, political, and economic problems of waging a world war in the 1990s. Army officers study a map of western Europe that is spread out on a Ping-Pong-sized table, looking for reserves to counter a new Red offensive thrust toward the Channel. Air force officers consult smaller charts and microcomputer data bases as they plan strike packages for the next day's operations. In over twenty command centers, navy officers consult their status boards and their alphanumeric and graphics computer displays as they transmit their orders to widely dispersed fleet units. Downstairs, the large game floor of the Naval Warfare Gaming System hums with activity as the communications from each command center reach the duplicate terminals where umpires translate orders into action.

These are all examples of how the U.S. Navy, and the civilian organizations that support it, use wargaming to help identify and solve the many problems facing today's military forces. From orienting new operations analysts to the complexities of modern naval warfare with the help of a hobby wargame or helping junior officers try out some of the tactics they may one day be called upon to execute, to sorting out the procedures for integrating forces in joint and combined operations or exploring the global ramifications of maritime strategy, the U.S. Navy has been putting the principles of the previous chapters into operation as wargaming becomes an increasingly important tool in its efforts to understand its current capabilities and limitations, and to identify its future opportunities and obstacles.

The navy's links to wargaming's past and its leadership in **295** wargaming's present make it an informative guide to wargaming's future. Current navy uses of wargames include both educational and research purposes, and employ all of the techniques described earlier, including commercially available hobby board games, microcomputer games, seminar games, and games supported by large mainframe computer systems.

EDUCATIONAL USES OF WARGAMING

During the early 1980s, one of the navy's principal independent operations research centers, as part of a project in support of navy wargaming research, used several commercial hobby board games to educate and train new analysts. The game *Flat Top* (published by The Avalon Hill Game Company) was used to educate these analysts about the origins of U.S. aircraft carrier tactical doctrine in World War II. The game *Sixth Fleet* (published by Victory Games, Inc.) brought them forward into the modern era at the same level of action (theater). Moving up to the realm of global maritime strategy, the game *Seapower and the State* (published by Simulations Canada) allowed them to explore broader issues. Finally, to help them put maritime power into the perspective of a NATO–Warsaw Pact war, they were introduced to the problems of ground warfare on NATO's Central Front with the help of *Third World War: Battle for Germany* (published by Game Designers' Workshop). The use of these hobby games not only facilitated the orientation of these new analysts (whose backgrounds were in chemistry and industrial administration) to military problems, but also helped them in their subsequent analysis of navy wargames.

At the Surface Warfare Officer School in Newport, and at the Naval Academy in Annapolis (as well as other locations, including on-board ship), naval officers and midshipmen can study the fundamentals of tactics using a microcomputer gaming system known as the Naval Tactical Game. Better known as NAVTAG, the game was "designed to be interesting and enjoyable for the participants. This was done deliberately to encourage shipboard officers whose time is at a premium to use the system both as the keystone of a locally structured training program and as an almost [!] recreational device during off-duty hours."[1]

NAVTAG was originally designed as a manual game by Lieutenant Commander Neil Byrne, USN, and introduced in

296 1978. The updated, computerized NAVTAG uses three video displays supplemented by a hard-copy printer. Each player and the game director have a video display. The game is played in turns, each of which represent one minute of real time. During a turn, "each Player will assess his situation after reviewing selectable status reports and receiving information from a geographic plot. Additionally, model disclosures and intelligence notes may be provided by the Game Director. The Game Director assesses the overall situation, monitors the display of Player commands and their consequences, and notes those events, actions, and situations upon which he will comment during the postgame critique."[2]

The game focuses on the very tactical level, allowing players to command individual ships or entire battle groups. Tactical systems and capabilities are modeled in great detail, drawing on not only mathematical analyses, but also on the results of systems tests and fleet exercises. Although not claimed as a panacea, many, including Admiral Thomas Hayward as CNO, saw NAVTAG as an important weapon in the arsenal of educational devices needed to help navy officers keep their tactical skills up-to-date.

Seminar-style games, a mainstay of navy research gaming, have become increasingly popular in the educational realm as well. These types of games are relatively free-wheeling and flexible and have become a widespread and popular forum for facilitating the exchange of information and ideas.

In a recent application of gaming techniques to the education and training of operational commanders and staffs "on the job" as opposed to being in formal schools, battle-force staffs conduct imaginary operations from their ships while tied up in port. Such exercises employ the facilities, data systems, and even communications that would be used in an actual situation. This Battle Force In-port Training concept is a navy version of what the army and air force are trying to accomplish in the Warrior Preparation Center.[3]

The Naval War College is still the navy's principal center for educational gaming. As part of their annual schedule of nearly fifty games, the War Gaming Department usually conducts about a dozen games in support of courses given at the college. In addition, several games are run in cooperation with the other service

war colleges, and the naval war colleges of various South Ameri- **297**
can and European navies. The remainder of the Naval War College's games are conducted to support the various fleet commanders in chief, subordinate commanders, the coast guard, the joint
chiefs of staff, the chief of naval operations, and others. Many of
these games are educational in spirit, but most also contain many
of the critical elements of research games.

RESEARCH USES OF WARGAMES

Unfortunately, the details of most navy research games are classified and may not be discussed here. It is possible, however, to
give a general description of some of the navy's principal uses of
research gaming.

Exploring Operational and Tactical Issues

The Central Front/Maritime Option wargame was a seminar
game conducted at the Naval War College. During this game,
the navy explored several types of operations that could be used to
support the defense of NATO's Central Front directly in the event
of a major war. Most of the play of the game took the form of
extended discussions of situations and options. Two sets of opposing teams were used to explore possible phases of the war in two
different ways. One team was constrained at the beginning of
each phase by the results of the preceding phase, in the manner
typical of most wargames. The second team was not so constrained; they were encouraged to examine each phase of the war
in light of what naval forces could contribute under the most
favorable of circumstances. The discussions of the possible courses
a Central Front war might take, and of the role of maritime forces
in each case, helped to identify important insights into several
operational, tactical, and logistical issues involved in such navy
support. Game analysts synthesized many of those insights into a
list of potential options for the use of maritime forces to affect the
course and outcome of the fighting ashore.

For over fifteen years, the Atlantic Fleet has conducted a series of wargames as part of its Tactical Command Readiness Program. These games are designed to exercise the commanders and
staffs of various components of the fleet. The thirty-first game in
the series used the Naval Warfare Gaming System at the Naval
War College to explore the prospective fleet concept of operations

298 alluded to at the end of the previous chapter. Players of the game included the commander of the Second Fleet and many of his subordinate battle group and supporting commanders. The game play was centered in the Norwegian Sea during a period of rising tensions, which ultimately led to the outbreak of global war. The game provided the players useful insights into the importance of mutual support between naval and land-based forces. It did not, however, allow complete or balanced judgments about whether or not the overall effect of the operation was worth the risks entailed in carrying it out. It also paved the way for further exploration of the feasibility of executing the concept through the conduct of actual at-sea exercises, and the running of several analytical studies.

Investigating Program or Systems Issues

In addition to exploring such operational and tactical issues, the navy has broadened wargaming's application to investigating specific program- and system-related issues. The navy office for Tactical Employment of National Capabilities conducted a computer-assisted seminar wargame to explore issues related to the navy's exploitation of space-based capabilities. The game examined the effects of the existence and potential use of anti-satellite weapons on the decision-making processes of navy operational commanders. The play and analysis of the game also provided insights into the contributions that space surveillance systems could make to fleet decision making.

The Program Objectives Memorandum War Game series is the most ambitious navy wargame exploring program and system issues. The Program Objectives Memorandum itself is an important document in the navy's Planning, Programming and Budgeting system. The game, while not an integral part of preparing that document, has had a certain amount of influence in pointing out particular navy programs of great importance, which the Program Objectives Memorandum usually addresses. The game itself typically looks at the entire spectrum of future navy systems in the three major theaters of a global war (the Atlantic, Pacific, and Mediterranean). The game uses a seminar format to identify and examine programmatic issues to improve the Naval Warfare Appraisal (another major document prepared in support of the navy's budget development process) and to highlight selected issues that may require additional study, analysis, and gaming in **299** the next year's cycle.

Developing Insight and Consensus about Strategic Issues

Just as the Program Objectives Memorandum wargame series is used by the navy's director of Naval Warfare as a catalyst for high-level discussions of warfare-related systems issues, other war-games serve as forums for the exploration of broader issues of strategy and policy. Each year since 1981, the chief of naval operations has appointed several senior navy commanders and captains, as well as marine corps officers of similar rank, to the Strategic Studies Group at the Naval War College. This group has a charter to investigate important issues in navy strategy and concepts of operations. Wargaming is an important tool in the Strategic Studies Group's efforts to refine and develop the navy's Maritime Strategy. The chief of naval operations himself has employed wargaming directly in his conferences with the navy commanders in chief to help build a consensus among the highest navy leadership about the strategic direction the service should pursue.

The largest and most visible of the navy's strategic-level games is the Global War Game series, held each summer at the Naval War College. The series is concluding its tenth year and its second cycle of five games. The last five games explored the capabilities of the United States and its allies and enemies in a global war set in 1990. Particular emphasis in the Global series is placed on identifying issues requiring further study and attention in the planning of global strategies and, in the most recent games, strategies for prolonged conventional war.

Because of the wide-ranging and complex objectives of the Global War Game series, the structure and format of the games has been similarly complex. Red and Blue teams have been headed by a National Command Authority, representing the civilian leadership of the Soviet Union or the United States. Military commanders have been played from the level of the Soviet General Staff or the U.S. Joint Chiefs down to the level of theater or fleet commanders. Play at the level of a navy battle group, army corps, or below has generally been handled by game controllers. In addition to the principal Red and Blue teams, a Green team has played all nations other than the U.S. and USSR, and

300 has also represented international organizations like the United Nations.

In the Global War Game, each game day represents about two days of actual war-fighting time. The three-week duration of the game limits the number of days of war that can be played in a single year, preventing the exploration of important issues of a long war. To get around this problem, the designer of the current five-year series of games, Mr. O.E. "Bud" Hay, adopted an approach that linked the last four games of the series together. With minor modifications, the ending situation of one game became the opening situation for the following year's game. Despite the potential difficulties of such an approach over a prolonged period of time, the game has remained a successful and influential one.

The nature of the Global War Game series has made it an important and effective device for helping the navy explore its own strategy in the context of the national military strategy, and in concert with the concepts of its sister services and allies. The participation of experts in fields ranging from international monetary policy to highly advanced technologies have helped to make the Global series especially rich in the number and types of issues it can raise. In fact, in early 1987, Mr. Hay briefed insights from the Global War Game series to the Senate Armed Services Committee as part of their review of the development of U.S. military strategy and policy.

SUMMARY

The previous chapter argued that the future of professional wargaming must be based on the integration of wargames, exercises, and analysis, along with historical insights and operational experience, into a coherent approach to the study of current and prospective military problems and opportunities. The United States Navy, in its continuing efforts to use wargaming as a forum, a classroom, and a laboratory, has played the principal role in laying the groundwork for such integration. In its attempts to build a complete and balanced understanding of future maritime realities through an ongoing process of modeling, testing, trying, and evaluating, the navy reveals the path for others to follow.

11

WHAT OF THE FUTURE?

But where does that path lead? Even more importantly, where should it lead? Part I showed that both hobby wargaming and professional wargaming stand poised on the brink of another serious downturn in their stature and popularity. Part II outlined some guiding principles of wargaming in hopes of helping both hobby and professional wargamers reverse that trend and at last smooth out the up-and-down cycle that has plagued wargaming for so long. The time seems ripe. Never before have wargamers enjoyed such a wide variety of powerful and effective tools and techniques. Yet never before has such an overabundance of technology so threatened to fragment the community of wargamers and frustrate their fondest hopes. Wargaming's delicate balance once again threatens to topple.

A QUESTION OF BALANCE

Wargamers have always struggled to maintain a balance between realism and playability. Some gamers believe the source of the problem is the supposedly inherent tension between those two facets of wargaming's dual nature, between detail and simplicity.

302 They are, at best, only partially correct. Realism is not the same thing as minute technical detail, nor is playability equivalent to abstraction or simplicity.

One reason for the debate between those who emphasize realism and those who stress playability is that they are really disagreeing about something else, the balance between the game system and the game player. Too often game designers fail to consider the player and his role adequately. Instead, they try to produce realism by designing a game system that explicitly accounts for everything that could physically affect a situation. That is an admirable goal, but only up to a point. The critical point is reached when making the system work demands too much artificial behavior from the player. Each added artificiality distorts the player's perceptions of and psychological reactions to the very reality that the game is trying to re-create. Tactical-level games, for example, cross the line when players must spend an hour or more deciding how each individual unit will operate, totaling terrain modifiers, and calculating combat odds in order to represent a thirty-second span of real time.

Realistic games allow players to behave naturally. Intricate, specialized, unnatural systems make the players more concerned about keeping track of rules and carrying out mechanical operations than experiencing the realities of decision making. On the other hand, overly simplistic, artificial systems, which justify "fudging" realism with the convenient excuse of increasing playability, rob the player of the opportunity to make decisions on the basis of his or her own perception of the critical elements of a real situation.

This relationship between what goes on in the mechanics of the game and what the player actually experiences during play is one that too often does not receive enough attention. What, after all, is the point of a wargame if not to give its players as realistic an experience of military decision making as possible? Unfortunately, designers, developers, controllers, and others involved in creating and facilitating the play of wargames sometimes become too concerned about giving the customer what he wants or expects, even when those expectations may be historically or realistically inaccurate, or entail what can only be described as an illegitimate use of gaming.

These latter problems are reflected in the tendency of hobby **303** wargames to give the player too much control over his forces (the chess syndrome) and to exaggerate well-known effects so that players think the game is accurate because it corresponds with what they expect to see. Bill Nichols, designer of *Long Lance* and other games, was kind enough to point out this latter problem to me. His example was that every good wargamer knows that a German 88-mm antitank gun was very effective, much more so than a 76-mm weapon. His perception was that many designers overexaggerate the differences because gamers expect the "88" to be orders of magnitude better than anything else.

The same tendency can be seen in some professional games, in which too much stress is placed on numerical outcomes and measures. In many of its tactical games, the U.S. Army collects detailed data concerning "killer/victim" scores—that is, the number of each type of enemy unit killed by each system. How often do such scores become results used to proclaim "answers" rather than signposts that warn of problems or suggest topics for further exploration? Far too often, especially with the ever-present pressure from the customer to produce "statistically significant" proofs of one point or another.

The key to realistic wargaming lies in balancing the player's experience in his decision-making role with as accurate a representation as possible of the physical outcomes of his own decisions, his opponents decisions, and the objective dynamics of combat. Achieving this balance is difficult because realistic decision making requires giving the player realistic information and accurately representing the realistic effects of time.

Both information and time can be difficult factors to incorporate accurately into hobby wargames. Typically, the player has too much information available, of too high quality, and can spend too much time processing that information and arriving at a decision. As described above, this problem is particularly noticeable in tactical-level games. In higher-level games, the problem mutates into one of correctly scaling the pace of play to reproduce the appropriate lags in reaction time.

For example, gaming the entirety of World War II in the Pacific has always been a difficult task for hobby designers. That war was characterized by long months of relative inactivity followed

304 by short days and hours of intense and decisive carrier combat. Games using moves of many months (such as GDW's *Pearl Harbor* or Hobby Japan's *Pacific Fleet*) have difficulty capturing those dynamics, just as games using shorter time-steps (like SPI's venerable *USN*) make the inactive periods seem interminable. In his *Pacific War* game, Mark Herman achieved some success with a telescoping time scale to reflect the conditions more accurately.

Professional games have a significant advantage over hobby efforts at simulating the flow of information and time because they often have ready access to human umpires to oversee the necessary mechanics. Unfortunately, professional games have problems of their own. Umpires can limit the information they give to players and "muddy the waters" with false or inaccurate data. Umpires can also be arbitrary and inconsistent in their handling of such factors. Similarly, there is a strong tendency for human umpires to lapse into reporting information in game terms rather than real terms. In professional games, as in hobby games, the flow of time can be a problem because battle-damage assessment often employs complex mathematical models, which require game-control staffs to expend a great deal of time and effort to produce results.

Because wargames must so often rely on such mathematical models, the growing power of electronic computers, especially of small, portable ones, has been a boon to hobby and professional wargamers alike. In many ways, computers may hold the key to solving the problems of information and time, and thus may hold the key to balancing the player's experience and the game's dynamics.

Computers can store large quantities of detailed, multidimensional data easily, and can access and manipulate them in complex ways far more quickly than any human player or umpire. Computers can help simulate the unfolding of events continuously over time, placing players under more realistic pressure to decide in time. Computers can also be programmed to limit the information available to the players in a consistent and realistic way. The power of computer assistance for wargaming, however, can be pushed too far, particularly when the computer begins to replace the real intelligence, stupidity, and emotions of human players with the pure logic of artificial intelligence.

COMPUTERS AND ARTIFICIAL INTELLIGENCE: THE INHUMAN OPPONENT

One of the most prominent applications of artificial intelligence to wargaming can be found in the Rand Strategy Assessment System (RSAS). RSAS grew out of Andy Marshall's search for new and better techniques to assist in the process of net assessment. It was designed "to make wargaming more efficient, rigorous, and analytical. . . . [and] involves the use of artificial intelligence techniques to produce computer models able to replace some or all of the human teams. This speeds game play, allows [analysts] to examine *many* scenarios, and very importantly imposes a rigorous discipline requiring statements of assumptions and rationale."[1] This last point is, indeed, a very important one, and the real source of RSAS's value. It is an excellent tool for exploring the potential implications of assumptions.

The basic approach of the RSAS is built around four main ideas. The most obvious and well-known is the notion of "automated wargaming," in which "human players [are] complemented by or largely replaced by computer models acting as automatons or 'agents.'"[2] The key element in this approach is the use of rule-based models to represent the player-agents. The final two pieces of the puzzle are the use of campaign analysis and interactive force operations modeling to support the play of the game. In its fully automated mode, however, RSAS ceases to be a wargame and becomes merely another analytical computer model. Unfortunately, the principles of artificial intelligence that underlie the RSAS's philosophy are becoming more and more a part of gaming.

Even when the computer is used only to provide human players with a ready-made opponent, artificial intelligence has its dangers. Wargames are best played between human competitors, and the differences between the viewpoints, experience, and behavior of different players are what make wargaming such a rich source of insight. Artificial opponents in the shape of computer programs limit the human players of a game to contending with a single opponent, the designer of the computer program.

The danger is not in artificial intelligence itself, but in using it to replace the human player. The "computer opponent" is

306 valuable. It allows the human gamer to explore situations and possibilities on his own, to "practice," and gain directly the synthetic experience needed to play a game well. The mistake lies in playing only against the machine; that is a potentially misleading and eventually sterile practice.

Unfortunately, especially in the wargaming hobby, it is a practice that has taken hold. Hearsay and informal surveys (as well as the author's own personal experience) indicate that over 90 percent of the hobby wargames played on the computer are played solitaire. The dangers of such exclusive "playing against the system" should be clear. In many ways it hearkens back to the Naval War College's experience in the 1930s.

During those games, the "system" was so ingrained into the perspectives of the players that they found themselves unable to think originally enough, to ask enough of the right questions, and as a result the tactical games proved incredibly poor reflections of actual wartime experience in surface combat. Some of the NWC's errors are attributable to poor technical estimates—for example, the underestimates of the range, speed, and lethality of the Japanese Long Lance torpedo. Others were errors in the representation of the workings of command.

Interestingly, hobbyists are about to get a chance to experience a taste of the Naval War College's tactical gaming system of the 1930s in an as-yet-unpublished computer game, designed by Commander Alan Zimm, USN. A comparison of Zimm's original design and the work of Bill Nichols's games that deal with many of the same actions, is instructive of two directions in which both hobby and professional gaming may go in the future.

"YOU ARE IN COMMAND" OR "YOU ARE IN CONTROL": A CASE STUDY

In 1987, Simulations Canada published *Long Lance*, designed by William Nichols. The game presented players with tactical-level scenarios of the surface battles in the Solomons Islands region during 1942 and 1943, the very battles that showed the failures of the War College's tactical games. *Long Lance*, and its follow-on game *In Harm's Way*, present the players with "the viewpoint of the senior naval commander present for each side. As such, they are responsible for planning their ships' missions before the battle, and commanding their forces in action against

the enemy."[3] Players organize their forces into individual action **307**
groups, plot their initial course, and define the rules of engagement and battle doctrine (within strict limits) when the force contacts the enemy. Once the firing starts, the player must exercise
command from the confines of his flagship. His knowledge of the
conditions on other ships, and his direct control of their actions,
are restricted by the uncertain mercies of the radio.

In his designer's notes to *Long Lance*, Nichols outlines the
heart of his design problem and the soul of his solution: "The
command problems were so intimidating that at first I feared
the game could not be done. How could one simulate this lack of
control in a game, yet satisfy the players' proper desire to be more
than a mere bystander? . . . [The solution lay in] the importance
of battle doctrine and communications . . . in allowing the historical commanders to control their forces. . . . By giving the
player the ability to create his own doctrine, which can guide his
ships' actions when they are not under his direct control, the historical 'fog of war' could be preserved while giving the player a
challenging input. . . . By simulating the ship-to-ship radio, the
player is able to react to unforeseen events, but this is not a panacea."[4] Indeed, in the heat and confusion of battle, orders go un-
heeded, and the most precise plans and neatly arranged formations soon disintegrate. The result can be frustrating in the
extreme, but it seems to do a very good job of capturing the flavor
and confusion of the Solomons battles.

Commander Zimm's game (whose working title as of this
writing is *Action Stations!*) places far less stress on the problems
of fleet command. Instead, Zimm focuses on reproducing the
physical environment and dynamics of the engagement. Although the player does have overall command of his force, the
system also gives him the ability to control the maneuvers and
fire of each individual ship. As a result, the command, control,
and communication problems that lead to confusion and disarray
are lacking.

What is not lacking is a detailed re-creation of the complexities of naval warfare during that period. The game forces "a commander to make critical decisions constantly. How do I distribute
my fire? Should I lay smoke? Can I stay on a steady course to
optimize my gunnery performance, or is the torpedo threat too
high? Should I illuminate? How can I get to a torpedo launch

308 position without losing all my destroyers? Stack gasses are interfering with my fire—should I slow? Should I change course and present my armor at an angle to his fire for additional protection? What will that do to my closure rate? That cruiser's got a jammed rudder—should I abandon or protect it?"[5] The player thus faces both fleet-level and ship-level decisions. At the fleet-level, however, the player can make decisions with the assurance that his subordinates will carry out his intentions perfectly.

Although aware of the skewed view of fleet-level command this approach gives the player, Zimm was concerned that the lack of solid data about the performance of World War II navigation and communications systems, as well as the changing roles, importance, and capabilities of radar, would make any attempt to introduce those considerations difficult at best and mere guesses at worse. Instead, the player is presented with a picture of the tactical situation in a form similar to that employed by the Navy's NTDS (Naval Tactical Data Systems). The result is an interesting and challenging game of battles in all theaters and all phases of the war. It presents the players with a reasonable picture of the role of the ship's captain, but does not give as good a feeling for fleet command under conditions of limited control over subordinate units.

WHO WANTS ROLE SIMULATION ANYWAY?

Nichols terms his design style in *Long Lance* "viewpoint oriented." The more widely used hobby term is "role simulation." The role simulation places the game player in a well-defined role and attempts to re-create that role's decision-making responsibility. It also tries to restrict the information the player receives as well as limit his control by the C^3 limitations of his real-life counterpart.

The goals of the designers of role simulations are worthy ones, well in keeping with the value and principal purposes of wargaming. Unfortunately, role simulations have been received in the hobby with something less than overwhelming enthusiasm. When Frank Chadwick articulated the ideas of role simulation in the early 1980s, he was to some extent, at least, rebuffed by certain elements of the wargaming community. One gamer wrote: "Most games present much more than the view from a certain participant's point of view. . . . Nor do I feel that role-playing adequately covers the extensive amount of information **309** found in . . . games. . . . Granted, role-playing is a fine generalization to describe wargames to a beginner but it does not nearly define the things I expect to get out of any wargame I purchase."[6]

Today, there seems to be a definite split in the hobby, particularly among computer-game designers and players, between those who prefer "games" and those who favor "role simulators." The distinction, in the clearest hobby terms, is that "with a 'game' you move the pieces around, and in a 'role simulator,' you order them to move around themselves (something they may or may not do). . . . The differences between the two approaches have become so distinct . . . that one finds battle lines being drawn and heated discussions about which is the 'valid' use of a computer in wargaming."[7] Despite the perception of some within the hobby that "neither side is having trouble selling games, in other words, gamers want and appreciate the values of both approaches,"[8] Nichols believes that those who prefer to control all of their forces all of the time are in the majority, or, at least, that such games are more marketable. This, from one of the chief standard bearers of role simulation, may seem a distressing trend for the future.

Others, myself included, disagree with Nichols's pessimistic assessment. What the players really want, have always wanted, are games that are interesting, challenging, fun, and "good players." And, incidentally, they want some taste of history—some players more, some less—but nearly all will take as much as they can get without destroying the interest of the game as a game. The failure of the role simulators in capturing a greater share of the hearts and minds of the players, if indeed they have failed to any great extent, has been in their execution, not in their concept.

Professional games, not surprisingly, tend to be better at role simulation than hobby games. The professionals understand the nature of military command roles better than most hobbyists. They are also more sensitive to the nature, importance, and workings of the chain of command. The professionals, especially, understand the critical importance of communications, and of standard operating procedures for dealing with situations that occur when communications are interrupted. Unfortunately,

310 that awareness does not by itself guarantee better designs. Indeed, one of the most ubiquitous criticisms of the Naval Warfare Gaming System has been its poor representation of communication systems and employment. Professional gamers, as well as hobby gamers, must work harder to re-create the commander's role and point of view.

Craig Besinque, a hobby board-game designer, outlined some of the key principles of role simulation.[9]

- *Obscure the situation* by concealing as much as possible about unit capabilities, unit locations, future events such as the arrival of reinforcements or supplies, enemy capabilities, and enemy intentions as defined by specific victory conditions.

- *Obscure the mechanism* by limiting the player's ability to predict precisely the likely outcome of combat or other operations, and by rewarding rational decisions in the face of such uncertainty.

- *Simulate command pressures* by forcing players to make compromises between the speed of their reactions to events and the quality of those reactions, by creating doubt in the player's mind, and by designing the game so that players may attempt to understand and manipulate the psychology of their opponents.

- *Simulate command effectiveness* by limiting the span of control of the player and his subordinates (as represented in the game system), as well as allowing for varying degrees of determination and intensity in the performance of combat units.

- *Simulate command psychology* by accurately representing the capabilities and limitations of the player's forces and position, allowing him to experience some of the pressures caused by confidence or doubt about the ability of his force to handle the enemy successfully.

- *Simulate the scope of options* by freeing the player's choices from the restrictions of historical or doctrinal hindsight.

The use of those principles can sometimes be carried to extremes, especially when designers so obscure the situation and system that the player can have no synthetic experience by which to guide his actions. Nevertheless, they succinctly summarize some of the most important areas of required research and experimentation for future wargamimg.

Relatively few games in existence today have succeeded in following the principles listed above. (Besinque's own *Rommel in the Desert* is one of the more effective ones, and it is a board game!) Part of the difficulty, especially for board-game design, is **311** the almost certain need for a live opponent. Manual gaming poses almost insuperable difficulties, especially in the psychological elements of command.

Another problem is the difficulty and possible arbitrariness that may enter into techniques for preventing the player from exerting too much control over his forces. In *Panzer Armee Afrika*, for example, Jim Dunnigan included a "panic" provision that prohibited a player from moving any units in hexes whose grid numbers ended in a specific set of digits, which were chosen by a die roll. This problem is not insoluble, however, and even solitaire-board-game designers have come up with clever concepts for making things work at least reasonably well. (Jon Southard's recent *Tokyo Express*, a board game of the Solomons naval battles, is an excellent case in point.)

Computers again offer an important addition to the tools of role simulation. Computers can easily restrict the player's information or provide inaccurate or imprecise information. They can be programmed to represent the player's subordinates or even his superiors. Here is where the true value of artificial intelligence lies in wargaming, especially when it is carefully linked to a system of "doctrine definition" that allows the player to "train" or prepare those subordinates to act as much according to his wishes as is reasonable.

The danger, in the hobby at least, is that designers, publishers, and distributors will yield too easily to the pressures of a marketplace somewhat wary of the relatively unsophisticated higher-level role simulations currently available. An overwhelming emphasis on the "electronic board-game" approach may stifle the creativity of the very designers who could make role simulators excellent games also. In some ways, the role-simulation approach to wargame design faces the same problems Charles Roberts did when he designed *Tactics*. It presents players with a new way of playing, a new way of thinking, one that prospective players will have to adapt to and learn how to play properly. Yet if Roberts could create a hobby almost out of thin air, it seems little enough to ask today's designers to help shape that hobby's future course in valuable directions.

The popularity of "first-person" games shows that the market potential for role simulation exists. Games like *Silent Service* and the more up-to-date *Red Storm Rising* and *Battlehawks* (a

312 high-resolution color-graphics game of air-to-air combat in the carrier battles of the South Pacific during World War II) show that role simulation can work. The key to these games is their representation of what the player can reasonably expect to control. They give the player a chance to practice and gain synthetic experience. When something goes wrong, the player can usually understand why and sometimes can take action to correct it. To a very great degree, the player's fate is in his own hands.

Re-creating higher-level roles requires creating the same options for investigation and correction while reducing the player's complete control. Adapting some of the ideas of first-person games, especially the personal element of a true player's point of view, is an as yet unexplored possibility in higher-level games. With increasing computer power and sophistication it may become possible to combine the detailed representation of physical events seen in Zimm's *Action Stations!* with the command perspective of Nichols's *Long Lance*, supplemented by the personal, and sometimes visual, perspective of a *Red Storm Rising*. Such a game would truly represent the best of all worlds.

NETWORKING AND DISTRIBUTED PROCESSING

Another potentially valuable future direction for wargaming lies in networking, tying together multiple computers, possibly from separate remote sites, to play simultaneously in a single game. Not surprisingly, the professionals, with their greater facilities and funding, lead the way in the use of networking and distributed processing for wargaming.

One of the most impressive examples of the approach is JANUS, an interactive computer wargaming system developed and managed under the aegis of the U.S. Army's Conflict Simulation Center at Lawrence Livermore National Laboratory. JANUS is a two-sided wargaming/analysis system, hence the name. (Janus was the Roman god of portals with a face on the front and back of his head allowing him to look both inside and outside the home at the same time.)

JANUS was designed to allow analysts to explore the effects of individual weapons systems in some detail, but in a more realistic, dynamic environment than that provided by purely analytical models. The emphasis on detail requires the user to specify three-dimensional terrain data, ground cover, and other special features. Other basic inputs are the characteristics of each system **313** and vehicle (including details like vehicle dimensions). Nuclear, chemical, and even futuristic "beam weapons" can be simulated in addition to the more usual conventional types.

JANUS uses "high-resolution graphics, intelligent work stations, distributed processing, and real time scenarios [to help] analysts . . . look at the effects of new weapons and tactics. Commanders and their staffs wanted to be able to refine the decision-making process while developing and testing realistic operations plans."[10] JANUS gives its players a continuous, map-like display of their forces and those of the enemy they have contacted or about which they have intelligence information. In addition, the player can access detailed status reports about the current strength and missions of his units and the status of the logistical situation.

The system can handle up to eight players on each side. The players may control up to 8,000 individual military sub-units, ranging from infantry fire teams, to tanks, to helicopters. The system can represent combat between a Blue brigade-level unit and a Red division-sized force. The players command and control their forces using the system's interactive graphics features, and the computer resolves activity and combat using an event-sequenced stochastic simulation, which resides in a VAX mini-mainframe computer. The players interact with the simulation in near-real-time through the use of as many as eight graphics workstations. These "intelligent workstations" preprocess the players' graphically entered commands to feed the main simulation. They also translate the results and display them to the players, thus allowing the main computer to concentrate on processing the simulation. This distributed approach combines the ease of graphical input for the players with the speed of a dedicated processor for simulation.

Despite its popularity and success as both an analytical tool and a training device, JANUS is limited in the types of problems it can address. It currently does not include models for tactical air support (beyond helicopters) or explicit C^3I systems. Its design began in 1978 and so the system is behind the state of the art in some ways, especially in its ability to assist the player in controlling the multitude of individual elements he must command. It does, however, hold out the promise of better things to come in real-time or near-real-time, interactive, networked wargaming.

314   Recently the hobby has also begun to take its first faltering steps in the direction of multi-machine games. One of the first and currently most impressive examples of hobby computer games to allow simultaneous play is Panther Games's *Fire Brigade*. *Fire Brigade* is a ground-battle game set in the Soviet-German war at the end of 1943. Two players, one representing the commander of the German Fourth Panzer Army and the other the commander of the Soviet First Ukrainian Front, can connect their IBM-PC or Apple Macintosh computers through a modem or direct cable link and give their orders and monitor results simultaneously.

Computer networking holds out the promise of restoring the balance between the computer as a tool and aid for the player, and the computer as the inhuman opponent. Players can use systems like *Fire Brigade* to play solitaire against the computer for practice and to explore options. When linked to other machines, possibly hundreds of miles away, systems like JANUS allow the computer to enhance the experience of playing against a live, human opponent. Again, the best of both worlds may soon be within the wargamer's grasp.

THE FUTURE OF MANUAL AND SEMINAR-STYLE WARGAMES

The practical as well as philosophical need for human opposition remains critical in manual wargaming, even when some computer assistance is available. This will remain important in the future because there are still some things that manual games (and professional seminar games) do better than strictly electronic ones.

Manual games seem to be particularly effective at representing the higher-level operational and strategic situations. The player's ability to see physically the entire playing surface at once and to make visual judgments of the overall situation is still difficult or impossible for most computers to provide at such high levels of play.

Manual games have other advantages as well. In the hobby they are often more appealing physically and aesthetically. On a more practical level, handling the counters and seeing the map in its entirety often can help the player feel that he has a better grasp of the overall situation and that he understands what he is trying to accomplish.

Manual games are also more direct in their person-to-person interaction. For this reason, among others, the professional seminar game, either in a strictly manual mode or with computer assistance, remains a popular and powerful tool for both research and training purposes.

The problems with manual wargames today remain the classic ones. How does one reproduce enough of the physical reality without so overburdening the player with game artificialities that his experience of play only vaguely resembles real-life command? Unfortunately, there are no easy answers. There are, however, some clever ideas that point the way toward improved solutions.

First, manual games must help develop synthetic experience in their players. They must teach the players the fundamental mechanics in easy and effective ways before burdening them with the more demanding tasks. This applies not only to individual games but, in the hobby at least, to the entire set of available games. It is far more difficult for a novice wargamer to become a true hobbyist today than it was twenty, or even ten, years ago. "To have a good chance to enter the hobby successfully these days, a beginner . . . needs help in choosing the right games, learning the rules, and learning play techniques and strategies."[11] In other words, he needs synthetic experience, provided either by the games themselves or, better yet, by the right combination of game and personal help from a veteran wargamer.

Secondly, designers must make a more concerted effort to produce what Mark Herman calls "natural" game systems. Natural systems allow the player to operate in as natural and free a manner as possible. Although a game must have some formal structure, that structure should seek to exploit the nature of the real-life situation represented in the game to simplify the player's task rather than try to impose an artificial structure on reality to simplify the designer's problem.

In *Gulf Strike*, for example, Herman devised a system for re-creating the complex interactions of air, sea, and land power in a modern setting by the simple, but virtually untried, expedient of allowing the player to conduct operations with units of any type at any time during his turn. In *Pacific War*, the telescoping time scales give a good feel for the changing focus of high-level commanders from considerations of grand strategy to the details of launching strikes. In *Trial of Strength*, designer Dave

316 O'Connor helped players master a wealth of detail at the strategic level by creating a comprehensive system that made use of another novel approach to reflecting the time required to conduct operations.

Such gaming devices, traditionally considered to fall into the realm of improvements in playability, are at last beginning to be recognized for the crucial role they play in creating realism. Indeed, the importance of the hobby-board-game designer's art of structuring understandable, flexible, and workable systems has led to a much greater involvement of experienced hobbyists with the defense professionals than Charles Roberts ever envisioned while he was trying to sell the U.S. Army *Game/Train*.

THE HOBBYIST AND THE PROFESSIONAL

It is quite impossible to list all of the wargaming hobbyists who are now working in defense-related activities. It is difficult enough to do so for just the major hobby "personalities" like Jim Dunnigan and Mark Herman. One of the earliest to "turn pro" was Randall Reed, head of Avalon Hill's research and development staff in the late 1970s, who left the hobby to work on wargaming for the U.S. Marine Corps. James Dunnigan is a well-known independent consultant, and Mark Herman is working on gaming and simulation for a major defense-consulting firm.

In fact, Herman and Victory Games were instrumental subcontractors in the development of a major gaming system for the Net Assessment section of the office of the secretary of defense and other clients, including the National Defense University. The game, the Strategic Analysis Simulation (SAS) was discussed briefly in chapter 3. It is a prime example of how the hobby and professional worlds can and should interact.

SAS and its companion Tactical Analysis Module (TAM) combine the professional's understanding of critical elements of reality with the hobbyist's talent for designing usable systems. The design of the system combined the professional's judgment and experience with the hobbyist's traditional questioning attitude and well-developed "BS-detector." As a result, SAS/TAM is a much better and more accurate system than it would have been otherwise. At the same time it is far easier to play and interpret than many other larger and less-elegant systems for global-level wargaming. Once again, it reflects the importance of a balanced view, this time between that of the hobbyist and that of the professional.

A VISION: OR IS IT JUST A GLEAM?

This chapter has tried to present a brief sketch of some of the more important or interesting elements of today's state of the art in wargaming. Of necessity, it has left out or glossed over much of interest and value. Similarly, the author has tried to avoid lapsing into either black despair about wargaming's uncertain future, or rosy prophecy of its brighter tomorrows. The bright tomorrows are possible, but it is going to take a lot of work by a lot of folks to achieve them.

The reasons for the recurring declines in wargaming's popularity and use are many and complex. Most of them can, however, be categorized into two major sets. First, games that had little resemblance to the real world and that could not help people address real problems quickly lost their appeal because they seemed merely poor abstractions of what they promised. Second, overly complex games with highly specialized rules and procedures proved too time-consuming to learn and too artificial in play to produce useful and realistic insights into the very problems their complexities were theoretically designed to address. When the balance is lost, so too is the game.

Where then is the secret to the balanced wargame? This book has spent an enormous number of words presenting an historical perspective on wargaming, outlining its guiding principles, and exploring its future prospects. After all that, the reader might expect that the author could answer such a simple question. He cannot. The only secret lies in the experience, imagination, and talent of the wargamer and in the quantity and quality of the data on which his game is based. Chapter 3 referred to the Israeli belief in Sun Tzu's philosophy that, of the three elements of war—people, weapons, and wisdom—wisdom is the most important. So it is with wargaming. So must it be with wargamers.

HOT WASH-UP

The game is over. The hot wash-up begins. The players, control- **319**
lers, observers, and analysts gather together one last time before
they return to their duty stations and work-a-day jobs. They try to
determine just what actually happened during their bloodless
conflict and struggle to discern what it all might mean for the pre-
vention or efficient termination of a future "real thing."

Wargaming's tools have progressed enormously over the
centuries, from the smooth stones and simple wooden board of
Wei Hai to the high-resolution color graphics of JANUS. Yet the
origins of the gaming instinct and the motivations of the gamer
have changed surprisingly little. The wargame gives us a chance
to play at one of the most terrible creations of the hand of man, a
creation capable of destroying all we have ever built and, at the
last, the very world on which we live. It also gives us a chance to
reduce the terrifying dimensions of this monster of our own
making to a more human scale. More importantly, it gives us a
chance to reduce some of our terror of the beast long enough to
bring to bear against it the one weapon that, ultimately, is our
final and only defense—our reason.

320   In the millennium to come, perhaps mankind may at last find itself prepared "to study war no more." To hasten the arrival of that time, however, the exact opposite must be the case. For it is only by studying war, in all its real gore and false glory, by understanding its causes, its conditions, and its consequences, that men can find within themselves the weaknesses that lead to war's downward-spiraling staircase and the strength to avoid taking the first, fatal steps.

Wargames are not merely convenient aids and adjuncts to such a study. They are essential tools without which that study is doomed to risk potentially disastrous failure. The dry calculation of costs and benefits, relative casualties and destruction, of mere pencil-and-paper analyses of nuclear war take on a different aspect when a human player must weigh the decision to "nuke the bastards," even if it is only a game. I will never forget the somber atmosphere of a Global War Game when the U.S. president player discussed the prospects of nuclear escalation. It was only a game, but it made the chasm into which we all stare seem far too real.

The history of wargaming has revealed the strengths and limitations of the tool, both when it is wielded by the defense community as a means to shape strategy and policy, and when it is exercised by the amateur as a means to increase his understanding of the past, present, and future of war and peace. A wargame is not real, it is not repeatable, it is not a panacea. It is an exercise in human interaction and the interplay of competing wills. Its strengths lie in its ability to explore the role and potential effects of human decisions on the human ability to make war. Its weaknesses lie in its inability to re-create the actual physical conditions under which such decisions must be made or the actual consequences of those decisions. Then again, perhaps those too are strengths.

Over the years, the recurring failure of both wargamers and decision makers to understand the utility and the limitations of their tool has led to periods of great and unwarranted popularity, followed by periods of equally great and equally unwarranted disfavor. These wave-like cycles have affected both the professional wargame and the hobby wargame. In recent years we have experienced the up side of wargaming's manic-depressive tendencies.

But just as Andrew Wilson's 1969 book *The Bomb and the Computer* heralded the last great wargaming depression, Thomas Allen's *War Games* seems to present once again the specter of wargaming's imminent decline.

This must not be. It is time, far past time, to put an end to the futile and destructive cycle of wargaming's rise and fall and rise and fall and . . . Whatever this book's author may be, prophet or fool, sage or charlatan, or simply a wargamer, he does not delude himself into imagining that his book alone will solve the problem and calm the waves. No one book, no one wargamer, can do that. But the entire community, the wargaming professional and wargaming hobbyist, the operations researcher and systems analyst, the pundit and soldier, sailor and scholar, the entire community just might be able to pull it off.

War is too important to ignore and too deadly to misunderstand. Temporary protection, if not eventual salvation, lies in achieving a balanced understanding of what war is, how it is waged, and what it is all about. In the end, the key to achieving that balance lies in perceiving the implications of a simple truth about all wars: systems and weapons destroy things and kill people; man wages war. And it is man who is, after all, the ultimate object of wargaming's legitimate explorations.

APPENDIX

QUESTIONS FOR WARGAME ANALYSIS

PREPARATION

- What information was provided to participants prior to their arrival?
- How are game objectives defined in preliminary briefings?
- What information is briefed to participants before play begins?
- How and to what level of detail is the scenario described? What is it?

STRUCTURE AND STYLE

- What is overall game structure and style?
- Who are the players?
  - —Is there a team structure?
  - —From what commands do team members and leaders come?
  - —What are the names and real-world jobs of the principal players?
  - —How many sides are there in the game (one, two, or many)?
  - —What are the decision levels of the players, and how do they communicate?
  - —What are the responsibilities and limitations of the players, and how do these correspond to their roles?

324 ● What are the roles of control?
—How are command levels above and below the players represented?
—How do players and controllers/umpires communicate?
—What are controller/umpire responsibilities, powers, limitations?
● What is the formal analysis plan?
—How many analysts are there, and where are they assigned?
—What are analysts told to look for?
—What other instructions are the analysts given?
—Who has overall responsibility for analysis?
—How frequently do analysts meet?
—What are topics of discussion at analyst meetings?
—How will analysis be integrated, when, where, and by whom?

PLAY

● What data and displays are available to the players?
—What information is provided?
—What types of displays are employed (books, charts, computers)?
—What are the sources of the information?
—Are there any questions about the accuracy of the data?
—Are the data available and the players' access to it appropriate for the command level they represent?
—Is the detail of data available commensurate with its importance or merely driven by availability?
—How often, easily, and well do the players make use of the data displays? For what reasons?
● During the course of play, what decisions are made by the players, and which are left up to others (control, umpires, etc.)?
—How detailed are decisions regarding force employment?
—What sort of control do players have in combat situations?
—How well do players control reconnaissance and intelligence assets?
—Are players' questions focused on what they should do, what they can do, what they must do, what they will do, or how can they do?
● How are game events defined?
—What are players told about what is happening and when?
—What do control and the umpires *not* tell the players?

- How are events sequenced?
  —What defines a move (time, activity, other)?
  —How is game time controlled relative to real time (steps or clock speed)?
  —How do players' decisions construct sequences of events?
  —What is the level of player interaction and response to developing situations?
- How does battle-damage assessment (BDA) or event resolution work?
  —Who does BDA? When?
  —What techniques, models, data do they use and how?
  —How do they receive instructions and information about events to resolve? From whom? When?
  —What are the factors critical to individual resolutions or classes of actions?
  —How do umpires/BDA translate player decisions into force movements, interactions, etc.?
  —How are players given BDA results? With what frequency, time delay, and accuracy of reporting?
  —Is the "fog of war" appropriate for player decision levels?
  —How does BDA affect later decisions?

ATTITUDES

- What are players' feelings about their roles and ability to influence events?
  —What is the source of those feelings?
  —What do the players see as the good points of the process?
  —What do they see as the problems?
  —What critical decisions did the players make? Why did they decide as they did?
  —What were the critical factors, understanding, and prejudices affecting decisions?
  —What special insights and ideas did the players bring to the game, and how has the play of the game affected them?
- What are the attitudes of control?
  —How does the sponsor feel about the course and value of the game?
  —Does this feeling change? What influences it?
  —How do controllers/umpires feel about their role, and how well they are carrying it out?

326 —How do attitudes of the sponsor and control group about the course of the game and its smoothness or value compare to the attitudes of those in the trenches? What appears to be the source of any disagreement in these attitudes?

NOTES

INTRODUCTION

1. *Team Yankee* Game Rules, p. 2.
2. "1987 F&M Readers Poll," *Fire & Movement*, No. 56, p. 11.
3. H. G. Wells, *Little Wars*, p. 97.
4. James F. Dunnigan, *The Complete Wargames Handbook*, p. 109.
5. Stephen B. Patrick, "The Rommel Syndrome or Wargames as a Snare and Delusion," *Moves*, No. 1, p. 4.
6. Redmond A. Simonsen, "Walter Mitty Strikes Back! or Just Call Me Erwin," *Moves*, No. 1, p. 5.

CHAPTER 1. THE BIRTH OF THE WARGAME

1. Alfred H. Hausrath, *Venture Simulation in War, Business, and Politics*, p. 3.
2. Abe Greenberg, Captain, U.S. Navy, "An Outline of Wargaming," *Naval War College Review*, p. 93.
3. Francis J. McHugh, *Fundamentals of War Gaming*, p. 2-1.
4. Farrand Sayre, *Map Maneuvers and Tactical Rides*, p. 6.
5. The description of Helwig's game is taken from McHugh, *Fundamentals of War Gaming*, p. 2-3.
6. Sayre, *Map Maneuvers*, p. 6.
7. McHugh, *Fundamentals of War Gaming*, p. 2-5.

8. John Clerk, *An Essay on Naval Tactics, Systematic and Historical.* Quoted in McHugh, *Fundamentals of War Gaming*, p. 2-42.

9. McHugh, *Fundamentals of War Gaming*, p. 1-78.

10. The discussion of Venturini's game is based largely on McHugh, *Fundamentals of War Gaming*, pp. 2-3–2-5.

11. Quoted in Sayre, *Map Maneuvers*, p. 6.

12. Ibid., p. 7.

13. See Hausrath, *Venture Simulation*, p. 5.

14. "Reisswitz the Elder," *Militair Wochenblatt*, No. 73, unpublished translation by William Leeson.

15. General der Infanterie z.D. Dannhauer, "Das Reisswitzsche Kriegsspiel von seinen Beginn bis zum Tode des Erfinders 1827," (The Reisswitz Wargame from the Beginning to the Death of Its Inventor, 1827), *Militair Wochenblatt*, No. 56, unpublished translation by William Leeson.

16. General von Muffling, "Anzeige" (Notice), *Militair Wochenblatt* (Berlin), No. 42, unpublished translation by William Leeson.

17. Lt. G.H.R.J. von Reisswitz, "Anzeige" (Notice), *Militair Wochenblatt* (Berlin), No. 42, unpublished translation by William Leeson.

18. See Sayre, *Map Maneuvers*, p. 10.

19. Ibid., p. 12.

20. Julius von Verdy du Vernois, *A Simplified War Game*, p. 8.

21. Ibid.

22. Ibid., pp. 8–10.

23. C. W. Raymond, Major, U.S. Army, *Kriegsspiel.* Quoted in Sayre, *Map Maneuvers*, p. 15.

24. Sayre, *Map Maneuvers*, p. 28.

25. Ibid., p. 29.

26. H. G. Wells, *Little Wars*, p. 110.

27. From *The Engineer*, vol. 86, 9 December 1890, p. 581. Quoted in Donald Featherstone, *Naval War Games*, pp. 145–46.

28. Fred T. Jane, "The Naval War Game," *U.S. Naval Institute Proceedings*, p. 661 and p. 602, respectively.

29. Ibid., p. 601.

30. Ibid.

31. Ibid.

32. Featherstone, *Naval War Games*, p. 147.

33. Jane, "Naval War Game," p. 602.

34. Featherstone, *Naval War Games*, p. 148.

35. Ibid.

36. A.P. Niblack, Lieutenant Commander, U.S. Navy, "The Jane Naval Wargame in the *Scientific American*," *U.S. Naval Institute Proceedings*.

37. Featherstone, *Naval War Games*, pp. 149–50.

38. Niblack, "Jane Naval Wargame," p. 581.

39. See Thomas B. Allen, *War Games*, p. 124.

40. Ibid.

41. Friedrich Helfferich, "From the Dawn of Wargaming: *Schlactenspiel* and *Wehrschach*," *Fire & Movement*, no. 48, p. 13.

42. Ibid.

43. Rudolf Hofmann, *War Games*.

44. Ibid., p. 7.

45. Generaloberst Franz Halder, in his foreword to Hofmann, *War Games*, pp. ix–x.

46. Ibid., p. x.

47. Hofmann, *War Games*, p. 7.

48. Ibid., p. 30.

49. Grand Admiral Karl Doenitz, *Memoirs, Ten Years and Twenty Days*, pp. 32–33.

50. Hofmann, *War Games*, p. 16.

51. Ibid., p. 20.

52. McHugh, *Fundamentals of War Gaming*, p. 2-18.

53. Louis Morton, "Japan's Decision for War (1941)," in Kent Roberts Greenfield, ed., *Command Decisions*, p. 69.

54. McHugh, *Fundamentals of War Gaming*, p. 2-18.

55. Morton, "Japan's Decision for War," p. 73.

56. Mitsuo Fuchida and Okumiya Masatake, *Midway, The Battle that Doomed Japan*, p. 52.

57. Ibid., pp. 95–97.

58. Quoted in Gordon Prange, *Miracle at Midway*, p. 36.

59. Hausrath, *Venture Simulation*, p. 31.

60. Ibid., p. 32.

61. Quoted in ibid., p. 32.

62. Sayre, *Map Maneuvers*, p. 24.

63. Ibid., p. 25.

64. Hausrath, *Venture Simulation*, p. 23.

65. See Sayre, *Map Maneuvers*, p. 21.

66. Quoted in Andrew Wilson, *The Bomb and the Computer*, p. 12.

67. Philip H. Colomb, "Le Duel ou Jeu de la Guerre Naval," *Revue Maritime et Coloniale*, and A. Colombo, "Giuoco di Guerra Navale," *Revista Marittima* (Roma).

68. Wilson, *The Bomb and Computer*, p. 26.

69. Ibid., pp. 26–27.

70. Ibid., p. 26.

71. Ibid., p. 32.

72. Hausrath, *Venture Simulation*, p. 33.

73. W. R. Livermore, Captain, U.S. Army, *The American Kriegsspiel*, p. 26.

74. William Chamberlaine, Major, U.S. Army, *Coast Artillery War Game*. Quoted in McHugh, *Fundamentals of War Gaming*, p. 2-34.

330   75. "The War Game and How it is Played," *Scientific American*, p. 470.

76. *The Solution to Map Problems.* Quoted in McHugh, *Fundamentals*, p. 2-37.

77. See McHugh, *Fundamentals of War Gaming*, p. 2-38.

78. General von Cochenhausen, "War Games for Battalion, Regiment, and Division," *Military Review*, Mar. 1941.

79. Quoted in McHugh, *Fundamentals of War Gaming*, p. 2-38.

80. Ninth United States Army Staff, *Conquer, The Story of the Ninth Army*, p. 141. Quoted in Hausrath, *Venture Simulation*, p. 34.

CHAPTER 2. WARGAMING AND THE U.S. NAVAL WAR COLLEGE

1. For a brief and fascinating account of this period, see "The Parable of the Ships at Sea," chapter 8 in Elting Morison, *From Know-How to Nowhere.*

2. *Report of the Secretary of War, 1890.* Quoted in Spector, *Professors of War*, p. 5.

3. Spector, *Professors of War*, p. 10.

4. For more on the history of the Naval War College, see John B. Hattendorf, B. Mitchell Simpson III, and John R. Wadleigh, *Sailors and Scholars: the Centennial History of the U.S. Naval War College.*

5. Anthony S. Nicolosi, "The Spirit of McCarty Little," U.S. Naval Institute *Proceedings*, p. 74.

6. See ibid. and Hattendorf, *Sailors and Scholars*, pp. 26–27 for descriptions of these exercises.

7. Nicolosi, "McCarty Little," p. 74.

8. A. Colombo, "Giuoco di Guerra Navale" (Naval War Game), *Revista Marittima* (Roma).

9. From *Harpers Weekly*, Feb. 1895, quoted in Francis J. McHugh, *Fundamentals of War Gaming*, p. 2-47.

10. Letter from Little to Luce, quoted in Spector, *Professors of War*, p. 74.

11. Quoted in Spector, *Professors of War*, p. 74.

12. McHugh, *Fundamentals of War Gaming*, p. 2-49.

13. War College Rules for 1902, quoted in Spector, *Professors of War*, p. 78.

14. Francis J. McHugh, "Gaming at the Naval War College," U.S. Naval Institute *Proceedings*, p. 50.

15. Francis J. McHugh, "Eighty Years of Wargaming," *Naval War College Review*, p. 89.

16. See Spector, *Professors of War*, and Hattendorf, *Sailors and Scholars*, for a full discussion of this subject.

17. James A. Barber, Jr., Commander, U.S. Navy, "The School of Naval Warfare," *Naval War College Review*, pp. 89–96.

18. W. McCarty Little, "The Strategic Naval War Game or Chart Maneuver," U.S. Naval Institute *Proceedings*, pp. 1228–1229.

19. Ibid., p. 1213.

20. Ibid., p. 1230.

21. See Spector, *Professors of War*, p. 81.

22. Nicolosi, "McCarty Little," p. 78.

23. Little, *Strategic Naval War Game*, p. 1223.

24. Quoted in McHugh, *Fundamentals of War Gaming*, p. 2-48.

25. McHugh, "Gaming at the Naval War College," p. 49.

26. Quoted in Hattendorf, *Sailors and Scholars*, p. 113.

27. McHugh, "Gaming at the Naval War College," p. 51.

28. William S. Sims, Admiral, U.S. Navy, in *World's Work*, Sep. 1923, quoted in McHugh, *Fundamentals of War Gaming*, pp. 2-53–2-54.

29. McHugh, "Gaming at the Naval War College," p. 51.

30. Ibid., pp. 50–51.

31. Hattendorf, *Sailors and Scholars*, p. 120.

32. See Michael Vlahos, "Wargaming, an Enforcer of Strategic Realism: 1919–1942," *Naval War College Review*.

33. See, for example, Hattendorf, *Sailors and Scholars*, p. 127, and Vlahos, "Wargaming," p. 7.

34. Chester W. Nimitz, 1923 Naval War College thesis, excerpted in *Naval War College Review*, Nov.–Dec. 1983, pp. 12–13.

35. Vlahos, "Wargaming," pp. 18–19.

36. Ibid., p. 19.

37. T. J. McKearney, Lieutenant Commander, U.S. Navy, *The Solomons Naval Campaign: A Paradigm for Surface Warships in Maritime Strategy*, p. 187.

38. Quoted in Hattendorf, *Sailors and Scholars*, p. 143.

39. McHugh, *Fundamentals of War Gaming*, p. 4-17.

40. McKearney, *Solomons Naval Campaign*, pp. 111–12.

41. Ibid., p. 114.

42. Ibid., p. 115.

43. Hattendorf, *Sailors and Scholars*, pp. 164–78.

44. McHugh, "Gaming at the Naval War College," p. 52.

45. Ibid.

46. Hattendorf, *Sailors and Scholars*, p. 187.

47. McHugh, "Gaming at the Naval War College," p. 53.

48. Ibid., pp. 52–53.

49. Ibid., p. 53.

50. Hattendorf, *Sailors and Scholars*, p. 237.

51. Ibid.

52. McHugh, *Fundamentals of War Gaming*, chapter 5.

53. Richard S. Brooks, Lieutenant Commander, U.S. Navy, "How It Works—The Navy Electronic Warfare Simulator," U.S. Naval Institute *Proceedings*, p. 147.

54. Ibid., p. 148.

55. McHugh, "Gaming at the Naval War College," pp. 54–55.

56. McHugh, *Fundamentals of War Gaming*, p. 5-31.

57. Brooks, "How It Works," p. 147.

58. McHugh, "Gaming at the Naval War College," p. 53.

59. McHugh, *Fundamentals of War Gaming*, pp. 5-32–5-33.

60. Ibid., pp. 5-33–5-35.

61. Office of the Chief of Naval Operations, Office of the Assistant for War Gaming Matters (OP-06C), *The Navy War Games Manual*, p. I-1.

62. Ibid., p. III-2.

63. McHugh, Francis J, "War Gaming and the Navy Electronic Warfare Simulator," *Naval War College Review*, p. 29.

64. "NEWS for the Fleet!" *Naval War College Review*, p. 2.

65. McHugh, "War Gaming and the Navy Electronic Warfare Simulator," p. 30.

66. Ibid., p. 31.

67. Ibid., pp. 31–32.

68. Ibid., pp. 30–31.

69. McHugh, *Fundamentals of War Gaming*, p. 5-49.

70. Abe Greenberg, Lieutenant Commander, U.S. Navy, "Wargaming: Third Generation," *Naval War College Review*, p. 72.

71. Ibid.

72. Ibid.

73. Ibid., p. 73.

74. Ibid., p. 74.

75. Ibid.

76. Hattendorf, *Sailors and Scholars*, p. 254.

77. Ibid., p. 264.

78. Ibid., p. 272.

79. Ibid., p. 276.

80. Ibid., p. 278.

81. Quoted in ibid., p. 287.

82. Ibid., p. 297.

83. Julian J. Le Bourgeois, Vice Admiral, U.S. Navy, "President's Notes," *Naval War College Review*, p. 2.

84. The description of the Naval Warfare Gaming System comes from the brochure *Naval Warfare Gaming System*, published by Computer Sciences Corporation. (No date given.)

85. The discussion of the Tactical Command Readiness Program is based largely on a conversation between the author and Mr. Erv Kapos, who was instrumental in the design of the program and its subsequent development.

86. Personal letter to the author from J. S. Hurlburt, Captain, U.S. Navy (Ret.).

87. Hattendorf, *Sailors and Scholars*, pp. 312–13.

88. Ibid., p. 313, and Robert J. Murray, "A Warfighting Perspective," U.S. Naval Institute *Proceedings*, p. 68.

89. Murray, "Warfighting Perspective," p. 68.

90. Ibid., p. 74.

91. Garry D. Brewer and Martin Shubik, *The War Game*, p. 125.   **333**
92. Murray, "Warfighting Perspective," p. 74.
93. Ibid.
94. Ibid., pp. 74–75.
95. W. F. McCauley, Rear Admiral, U.S. Navy, "Fight Smart," *Phalanx: Newsletter of Military Operations Research*, p. 10.
96. Hattendorf, *Sailors and Scholars*, p. 75.
97. Murray, "Warfighting Perspective," p. 75.
98. Ibid., p. 79.
99. McHugh, "Wargaming at the Naval War College," p. 55.

CHAPTER 3. WARGAMING AFTER THE WAR

1. Philip M. Morse, and George E. Kimball, *Methods of Operations Research*, p. 1.
2. Keith R. Tidman, *The Operations Evaluation Group*, pp. 66–67.
3. Ibid., p. 67.
4. Ibid.
5. Ibid., p. 68.
6. Alfred H. Hausrath, *Venture Simulation*, pp. 40–44.
7. Ibid., p. 40.
8. Ibid., p. 43.
9. Quoted in ibid., p. 42.
10. Ibid., p. 40.
11. For more on this subject, see: Garry D. Brewer and Martin Shubik, *The War Game*; Thomas B. Allen, *War Games*; Andrew Wilson, *The Bomb and the Computer*; Hausrath, *Venture Simulation*; and Tidman, *Operations Evaluation Group*.
12. Dr. Lincoln P. Bloomfield, "Political Gaming," U.S. Naval Institute *Proceedings*, p. 59.
13. Ibid.
14. Ibid., p. 55.
15. Ibid., p. 61.
16. Ibid.
17. Ibid.
18. Ibid.
19. Ibid., p. 62.
20. Ibid., p. 63.
21. Ibid., p. 64.
22. Hausrath, *Venture Simulation*, p. 265.
23. Robert Mandel, "Political Gaming and Foreign Policy Making During Crises," *World Politics*, p. 631.
24. W. M. Jones, "Getting Serious about Serious Games for Serious Problems," talk presented at the symposium Serious Games for Serious Questions, U.S. Department of State, 3–4 March 1988, Washington, D.C.
25. *The Avalon Hill Game Company History*, p. 3.

26. Ibid., p. 4.

27. The staff of *Strategy & Tactics* magazine, *Wargame Design*, pp. 10–11.

28. *The Avalon Hill Game Company History*, p. 4.

29. Ibid.

30. The staff of *Strategy & Tactics* magazine, *Wargame Design*, pp. 11–12.

31. Ibid., p. 13.

32. *The Avalon Hill Game Company History*, p. 4.

33. The staff of *Strategy & Tactics* magazine, *Wargame Design*, p. 13.

34. *The Avalon Hill Game Company History*, pp. 5–6.

35. *The Avalon Hill Game Company History*, p. 9.

36. Ibid., p. 11, and a personal letter to the author from James Dunnigan.

37. Ibid., p. 11.

38. Ibid., p. 12.

39. For a discussion of this topic, see John Prados, *Pentagon Games*, pp. 13 ff.

40. Ibid., p. 14.

41. Ibid., p. 61.

42. Allen, *War Games*, p. 194. The following discussion of the Sigma game series is based largely on Allen, pp. 193–208.

43. Ibid., p. 194.

44. From the Sigma II-64 game records, quoted in Allen, p. 200.

45. For a striking comparison of the game and reality, see Allen, p. 206.

46. Hausrath, *Venture Simulation*, thoroughly documents the contributions of the Research Analysis Corporation.

47. Ibid., p. 253.

48. Ibid.

49. Ibid., p. 254.

50. Ibid.

51. Wilson, *The Bomb and Computer*, p. 210.

52. Hausrath, *Venture Simulation*, p. 275.

53. Ibid., p. 276.

54. Ibid.

55. The staff of *Strategy & Tactics* magazine, *Wargame Design*, p. 17.

56. Ibid., p. 27.

57. "Designer's Notes," *Moves*, no. 23, p. 27.

58. Nicholas Palmer, *The Comprehensive Guide to Board Wargaming*, p. 126.

59. Taken from ibid., p. 187.

60. Rodger MacGowan, "Firing Line," *Fire & Movement*, p. 2.

61. The staff of *Strategy & Tactics* magazine, *Wargame Design*, p. 34.

62. Don Greenwood, *"The Russian Campaign:* Developer's **335** Notes," *Fire & Movement,* p. 30. Written in response to Richard F. De-Baun and Frank Aker, *"The Russian Campaign:* Battle Report," *Fire & Movement.*

63. Hal Hock, *"Tobruk* Designer's Notes," *Fire & Movement,* p. 28.

64. Mark Saha, "Close Up: *Tobruk," Fire & Movement.*

65. Allen, *War Games,* p. 295.

66. Ibid., p. 303.

67. For a description of other meetings between Dunnigan and Marshall, see ibid., pp. 93–95.

68. Richard D. Lawrence, Lieutenant General, U.S. Army, "Playing the Game: The Role of War Games and Simulations," *Defense/86,* p. 8.

69. Allen, *War Games,* p. 310.

70. Ibid.

71. U.S. Army War College, *McClintic Theater Model, Volume I: War Game Director's Manual,* p. 1.

72. Ibid., pp. i–ii.

73. Ibid., p. 2.

74. Ibid., p. 4.

75. Kurt Nordquest, "Sophisticated Jutland," *The Avalon Hill General,* p. 26.

76. Ibid.

77. Redmond Simonsen and Dave Robertson, "SpiBus: SPI's Newsletter of Microcomputer/Conflict Simulations Applications, nr. 0001," *Moves.*

78. Ian Chadwick, "SpiBus: Why I'm Really Buying a Microcomputer and What I'll Play When It Gets Here," *Moves.*

79. Ian Chadwick, "SpiBus: Why I'm Really Buying a Microcomputer and What I'll Play When It Gets Here, Part 2," *Moves,* p. 29.

80. Dr. Ed Bever, "Board Wargames and Computer Wargames: A Comparison," *Fire & Movement,* p. 52.

81. Ibid.

82. Kirby Arriola, "Carriers at War," *The Grenadier,* p. 49.

83. Ibid.

84. Reiner Huber, from the talk "Gaming and Simulation—A German Perspective," presented at the symposium Serious Games for Serious Questions, U.S. Department of State, 3–4 March 1988, Washington, D.C.

85. Ibid.

86. J. P. Wood, from the talk "Prospects for Land Combat Modelling," presented at the International Symposium on Improvements in Combat Modelling at RMCS Shrivenham, 4–7 Sep. 1974.

87. Dr. G. P. Armstrong, from the talk "Advantages of Manual Wargaming," presented at the International Symposium on Improvements in Combat Modelling at RMCS Shrivenham, 4–7 Sep. 1974.

88. The discussion of the Israeli gaming experience is based on a presentation made by Colonel Ya' Achov D. Herchal, IDF, at the symposium Serious Games for Serious Questions, U.S. Department of State, 3–4 March 1988, Washington, D.C.

89. Much of the discussion of Soviet gaming is based on John Sloan, Ali Jalali, Goublem Wardak, and Fred Gressler, *Soviet Style Wargames.*

90. Ibid., p. 3.
91. Ibid., p. 4.
92. Ibid.
93. Ibid., pp. 5–6.
94. Ibid., p. 6.
95. Ibid.
96. Ibid.

CHAPTER 4. THE NATURE OF WARGAMES

1. Quoted in Bill Nichols, "CRT," *The Grenadier,* p. 22.
2. H. G. Wells, *Little Wars,* p. 101.
3. Ibid.
4. Ibid.

CHAPTER 5. DESIGNING WARGAMES

1. James F. Dunnigan, *The Complete Wargames Handbook,* pp. 235–36.
2. See, for example: Carl H. Builder, *Toward a Calculus of Scenarios,* and Dr. Dale K. Pace, "Scenario Use in Naval System Design," *Naval Engineers Journal.*
3. *Third Reich* First Edition game rules, pp. 26 and 27.
4. Pace, "Naval System Design," p. 63.
5. Builder, *Calculus of Scenarios,* pp. 20–21.
6. Francis J. McHugh, *Fundamentals of War Gaming,* pp. 3-25–3-28.
7. Ibid., p. 3-11.
8. Ibid.
9. Ibid.

CHAPTER 6. DEVELOPING WARGAMES

1. "Designer's Notes," *Moves,* no. 4, pp. 1, 28.
2. H. G. Wells, *Little Wars,* p. 11.
3. Ibid., p. 12.
4. Ibid., p. 15.
5. Ibid., pp. 22–23.
6. Alan Emrich, "*Russia:* A Breakthrough or Just Another Steppe Closer?" *Fire & Movement,* pp. 48–55.
7. Ibid., p. 53.
8. *Center for Naval Analyses Annual Report 1986.* Quoting from

John Maynard Keynes, *The General Theory of Employment, Interest, and Money* (London: Macmillan and Co., Ltd., 1961), p. 298.

9. Ibid., p. 14.

10. Dr. Allan L. Gropman, "In Pursuit of the Holy Grail," *Phalanx: The Bulletin of Military Operations Research*, p. 1.

11. Ibid., p. 16.

12. Wayne P. Hughes, *Fleet Tactics, Theory and Practice*, p. 5.

13. J. P. Wood, from the talk "Prospects for Land Combat Modeling," presented at the International Symposium on Improvements in Combat Modelling at RMCS Shrivenham, 4–7 Sep. 1974.

14. Ibid.

15. Wayne P. Hughes, "Attention!" *Phalanx: The Bulletin of Military Operations Research*, p. 14.

16. Robert McQuie, *Historical Characteristics of Combat for Wargames (Benchmarks)*, p. iii.

17. Frank Chadwick, "My Two Cents," *The Grenadier*, vol. 2, no. 1, p. 28.

18. Quoted in Bill Stone, "Game Designers' Census," *The Grenadier*, p. 9.

19. John E. Hill, *Squad Leader*, First Edition Game Rules, p. 31.

20. Bill Haggart, *Napoleon's Last Triumph*, unpublished designer's notes, p. 3.

21. Ibid., p. 5.

22. Nicholas Palmer, *The Best of Board Wargaming*, p. 39.

23. Chadwick, "My Two Cents," p. 29.

24. James Euliss, Commander, U.S. Navy, "Wargaming at the U.S. Naval War College," *Naval Forces*, p. 103.

25. Ibid.

26. James F. Dunnigan, *The Complete Wargames Handbook*, p. 232.

CHAPTER 7. PLAYING WARGAMES

1. Frank Chadwick, "My Two Cents," *The Grenadier*, vol. 2, no. 1, p. 28.

2. Quoted in Frederick D. Thompson, *Beyond the War Game Mystique: Learning from War Games*, p. 6.

3. Frederick D. Thompson, "Beyond the War Game Mystique," U.S. Naval Institute *Proceedings*, p. 84.

4. Ibid., pp. 84–85.

5. Adapted from ibid., p. 85.

6. See, for example, Lloyd Hoffman, "Accurate Adversarial Play," *Phalanx: The Bulletin of Military Operations Research*.

CHAPTER 8. ANALYZING WARGAMES

1. Frederick D. Thompson, "Beyond the War Game Mystique," p. 95.

338   2. Randall C. Reed, "Game Reviewing Biases," *Fire & Movement*, p. 19.

3. Bill Haggart, "Consistency is the Hobgoblin of Little Minds," *Fire & Movement*, p. 17.

4. Frederick D. Thompson, "Did We Learn Anything from that Exercise? Could We?", *Naval War College Review* discusses some aspects of the problem.

CHAPTER 9. INTEGRATING WARGAMES WITH OPERATIONAL ANALYSIS AND EXERCISES

1. Phillip M. Morse and George E. Kimball, *Methods of Operations Research*, p. 1.

2. Operations Analysis Study Group, United States Naval Academy, *Naval Operations Analysis*.

3. See P.M.S. Blackett, *Studies of War: Nuclear and Conventional*.

4. Thomas B. Allen, *War Games*, pp. 257–62.

5. Alfred H. Hausrath, *Venture Simulation*, p. 92.

6. Samuel Eliot Morison, *History of United States Naval Operations in World War II, Vol. I, The Battle of the Atlantic*, p. 2.

CHAPTER 10. NAVY WARGAMING TODAY

1. *Naval Tactical Game [NAVTAG] Training System, Device 16H3A, Executive Summary*, p. 2.

2. Ibid.

3. Colonel John M. Vickery, "Training for Today and Tomorrow: The Warrior Preparation Center," *Journal of Electronic Defense*.

CHAPTER 11. WHAT OF THE FUTURE?

1. Paul K. Davis and James Winnefeld, *The Rand Strategy Assessment Center*, p. v.

2. Ibid., p. 3.

3. William Nichols, *Long Lance* game rules, p. 1.

4. Ibid., p. 5.

5. Alan Zimm, Commander, U.S. Navy, *Action Stations!* unpublished game rules.

6. Taken from a letter from John G. Alsen published in Frank Chadwick, "My Two Cents," *The Grenadier*, no. 6, pp. 8–9.

7. Dr. Jay C. Selover, "Firing Line," *Fire & Movement*, p. 8.

8. Ibid.

9. Craig Besinque, "Role Simulation," *Fire & Movement*.

10. Allan R. Vannoy, "JANUS: The U.S. Army's Tactical Warfare Computer Simulation System," *The Wargamer*, p. 51.

11. Jon Southard, "Introductory Games: Problems and Analysis," *Fire & Movement*, p. 14.

BIBLIOGRAPHY

BOOKS, MONOGRAPHS, AND RESEARCH PAPERS   339

Allen, Thomas B. *War Games: The Secret World of the Creators, Players, and Policy Makers Rehearsing World War III Today.* New York: McGraw Hill Book Company, 1987.

The Avalon Hill Game Company. *A History of the World's First and Largest Wargame Publisher: Silver Jubilee. Avalon Hill's First 25 Years in Review.* Baltimore, Maryland: The Avalon Hill Game Company, 1983.

Blackett, P.M.S. *Studies of War: Nuclear and Conventional.* New York: Hill and Wang, 1962.

Brewer, Garry D., and Martin Shubik. *The War Game: A Critique of Military Problem Solving.* Cambridge, Massachusetts: Harvard University Press, 1979.

Builder, Carl H. *Toward a Calculus of Scenarios.* Rand Note N-1855-DNA. Santa Monica, California: The Rand Corporation, 1983.

Center for Naval Analyses. *Annual Report 1986.* Alexandria, Virginia: The Center for Naval Analyses, 1987.

Chamberlaine, William, Major, U.S. Army. *Coast Artillery War Game.* 3rd. edition. Fort Monroe, Virginia: Coast Artillery School Press, 1914.

Clerk, John. *An Essay on Naval Tactics Systematic and Historical.* Edinburgh: Constable, 1790. (Reprinted in 1964 by University Microfilms, Inc., Ann Arbor, Michigan.)

340 Coyle, Harold B. *Team Yankee*. New York: Presidio Press, 1987.

Davis, Paul K. and James Winnefeld. *The Rand Strategy Assessment Center: An Overview and Interim Conclusions about Utility and Development Options*. Santa Monica, California: The Rand Corporation, 1983.

Doenitz, Grand Admiral Karl. *Memoirs, Ten Years and Twenty Days*. Trans. by R. H. Stevens in collaboration with David Woodward. London: Weidenfeld and Nicholson, 1959.

Dunnigan, James F. *The Complete Wargames Handbook: How to Play, Design, and Find Them*. New York: William Morrow and Company, Inc., 1980.

Featherstone, Donald F. *Naval War Games: Fighting Sea Battles with Model Ships*. London: Stanley & Paul Company, Ltd., 1965.

Fuchida, Mitsuo and Okumiya Masatake. *Midway, the Battle that Doomed Japan*. Clark Kawakami and Roger Pineau, eds. Annapolis, Maryland: U.S. Naval Institute Press, 1955.

Greenfield, Kent Roberts, ed. *Command Decisions*. Washington, D.C.: U.S. Department of the Army, Office of the Chief of Military History, no date.

Hattendorf, John B., B. Mitchell Simpson III, and John R. Wadleigh. *Sailors and Scholars; The Centennial History of the U.S. Naval War College*. Newport, Rhode Island: Naval War College Press, 1984.

Hausrath, Alfred H. *Venture Simulation in War, Business, and Politics*. New York: McGraw Hill Book Company, 1971.

Hofmann, Rudolf. *War Games*. Draft trans. by P. Luetzkendorf. Washington, D.C.: U.S. Department of the Army, Office of the Chief of Military History, MS P-094, 1952.

Hughes, Wayne P. *Fleet Tactics, Theory and Practice*. Annapolis, Maryland: Naval Institute Press, 1987.

Livermore, W. R., Captain, U.S. Army. *The American Kriegsspiel. A Game for Practicing the Art of War upon a Topographical Map*. Boston: Houghton, Mifflin and Company, 1882.

McHugh, Francis J. *Fundamentals of War Gaming*. 3rd ed. Newport, Rhode Island: U.S. Naval War College, 1966.

McKearney, T. J., Lieutenant Commander, U.S. Navy. *The Solomons Naval Campaign: A Paradigm for Surface Warships in Maritime Strategy*. Master's thesis, Naval Postgraduate School, Monterey, California, Sep. 1985.

McQuie, Robert. *Historical Characteristics of Combat for Wargames (Benchmarks)*. Bethesda, Maryland: U.S. Army Concepts Analysis Agency, Jul. 1988.

Morison, Elting E. *From Know-How to Nowhere*. New York: Basic Books, 1974.

Morison, Samuel Eliot. *History of United States Naval Operations in World War II, Vol. I, The Battle of the Atlantic*. Boston: Little, Brown & Company, 1947.

Morse, Philip M., and George E. Kimball. *Methods of Operations Research.* New York: John Wiley & Sons, Inc., 1951. **341**

Naval Tactical Game [NAVTAG] *Training System, Device 16H3A, Executive Summary.* Washington, D.C.: SYSCON Corporation, no date.

Ninth United States Army Staff. *Conquer, the Story of the Ninth Army.* Washington, D.C.: The Infantry Journal Press, 1947.

Office of the Chief of Naval Operations, Office of the Assistant for War Gaming Matters (OP-06C). *The Navy War Games Manual.* Washington, D.C., 1967.

Operations Analysis Study Group, United States Naval Academy. *Naval Operations Analysis.* 2nd ed. Annapolis, Maryland: Naval Institute Press, 1977.

Palmer, Nicholas. *The Best of Board Wargaming.* New York: Hippocrene Books, Inc., 1980.

————. *The Comprehensive Guide to Board Wargaming.* New York: Hippocrene Books, Inc., 1977.

Prados, John. *Pentagon Games: Wargames and the American Military.* New York: Harper & Row, Publishers, Inc., 1987.

Prange, Gordon W., with Donald M. Goldstein and Katherine V. Dillon. *Miracle at Midway.* New York: McGraw Hill Book Company, 1982.

Pratt, Fletcher. *Fletcher Pratt's Naval War Game.* New York: Harrison-Hilton Books, Inc., 1940.

Raymond, C. W., Major, U.S. Army. *Kriegsspiel.* Fort Monroe, Virginia: U.S. Artillery School, 1881.

Report of the Secretary of War, 1890. Washington, D.C.: U.S. Government Printing Office, 1891.

Sayre, Farrand, Major, U.S. Army. *Map Maneuvers and Tactical Rides.* 5th ed. Springfield, Massachusetts: Springfield Printing and Binding Company, 1912.

Sloan, John, Ali Jalali, Goublem Wardak, and Fred Gressler. *Soviet Style Wargames.* Greenwood Village, Colorado: Science Applications International Corporation, Jun. 1986.

The Solution to Map Problems. Fort Leavenworth, Kansas: General Service Schools, 1925.

Spector, Ronald. *Professors of War: The Naval War College and the Development of the Naval Profession.* Newport, Rhode Island: Naval War College Press, 1977.

The staff of *Strategy & Tactics* Magazine. *Wargame Design. The History, Production, and Use of Conflict Simulation Games.* New York: Simulations Publications, Incorporated, 1977.

Thompson, Frederick D. *Beyond the War Game Mystique: Learning from War Games.* Center for Naval Analyses Memorandum 83-0271, 1983.

Tidman, Keith R. *The Operations Evaluation Group: A History of*

342   *Naval Operations Analysis*. Annapolis, Maryland: Naval Institute Press, 1984.

Totten, Charles A. L., First Lieutenant, Fourth United States Artillery. *Strategos: A Series of American Games of War Based upon Military Principles*. New York: D. Appleton and Company, 1880.

U.S. Army War College. *McClintic Theater Model, Volume I: War Game Director's Manual*. Carlisle, Pennsylvania: U.S. Army War College, Aug. 1981.

Verdy du Vernois, Julius von. *A Simplified War Game*. Trans. from the French by Eben Swift. Kansas City, Missouri: Hudson-Kimberly Publishing Company, 1897.

Wells, H. G. *Little Wars*. Unabridged republication of the first edition published in London [by F. Palmer] in 1913. New York: Da Capo Press, 1977.

Wilson, Andrew. *The Bomb and the Computer: Wargaming from Ancient Chinese Mapboard to Atomic Computer*. New York: Delacorte Press, 1969.

ARTICLES

Arriola, Kirby. "Carriers at War." *The Grenadier*, no. 31, Feb. 1987.

Barber, James A., Jr., Commander, U.S. Navy. "The School of Naval Warfare." *Naval War College Review*, Apr. 1969.

Besinque, Craig. "Role Simulation." *Fire & Movement*, no. 55, Sep.–Oct. 1987.

Bever, Dr. Ed. "Board Wargames and Computer Wargames: A Comparison." *Fire & Movement*, no. 49, Jul.–Aug. 1980.

Bloomfield, Dr. Lincoln P. "Political Gaming." U.S. Naval Institute *Proceedings*, Sep. 1960.

Brooks, Richard S., Lieutenant Commander, U.S. Navy. "How It Works—The Navy Electronic Warfare Simulator." U.S. Naval Institute *Proceedings*, Sep. 1959.

Chadwick, Frank. "My Two Cents." *The Grenadier*, vol. 2, no. 1, Jan. 1979.

———. "My Two Cents." *The Grenadier*, no. 6, May 1979.

Chadwick, Ian. "SpiBus: Why I'm really Buying a Microcomputer and What I'll Play When It Gets Here." *Moves*, no. 55, Feb.–Mar. 1981.

———. "SpiBus: Why I'm really Buying a Microcomputer and What I'll Play When It Gets Here, Part 2." *Moves*, no. 56, Apr.–May 1981.

Cochenhausen, General von. "War Games for Battalion, Regiment, and Division." *Military Review*, Mar. 1941.

Colomb, Philip H. "Le Duel ou Jeu de la Guerre Naval" (The Duel or the Naval War Game). Trans. into Italian by L. Rivet, *Revue Maritime et Coloniale*, Fevrier (Feb.) 1881.

Colombo, A. "Giuoco di Guerra Navale" (Naval War Game). *Revista Marittima* (Roma), Dec. 1891.

Dannhauer, General der Infanterie z.D. "Das Reisswitzsche Kriegsspiel von seinen Beginn bis zum Tode des Erfinders 1827" (The Reiss-witz Wargame from the Beginning to the Death of Its Inventor, 1827). Unpublished trans. by William Leeson, *Militair Wochen-blatt*, no. 56, 1874.

DeBaun, Richard F. and Frank Aker. *"The Russian Campaign*: Battle Report." *Fire & Movement*, no. 5, Jan.–Feb. 1977.

"Designers Notes." *Moves*, no. 4, Aug. 1972.

"Designers Notes." *Moves*, no. 23, Oct.–Nov. 1975.

Emrich, Alan. *"Russia*: A Breakthrough or Just Another Steppe Closer?" *Fire & Movement*, no. 53, May–Jun. 1987.

Euliss, James, Commander, U.S. Navy. "Wargaming at the U.S. Naval War College." *Naval Forces*, vol. 6, no. 5, 1985.

Greenberg, Abe, Captain, U.S. Navy. "An Outline of Wargaming." *Naval War College Review*, Sep.–Oct. 1981.

———, Lieutenant Commander, U.S. Navy. "Wargaming: Third Generation." *Naval War College Review*, Mar.–Apr. 1975.

Greenwood, Don. *"The Russian Campaign*: Developer's Notes." *Fire & Movement*, no. 5, Jan.–Feb. 1977.

Gropman, Dr. Alan L. "In Pursuit of the Holy Grail." *Phalanx: The Bulletin of Military Operations Research*, Dec. 1987.

Haggart, Bill. "Consistency is the Hobgoblin of Little Minds." *Fire & Movement*, no. 42, Winter, 1984.

Helfferich, Friedrich. "From the Dawn of Wargaming: *Schlactenspiel* and *Wehrschach*." *Fire & Movement*, no. 48, May–Jun. 1986.

Hock, Hal. *"Tobruk* Designer's Notes." *Fire & Movement*, no. 1, 1976.

Hoffman, Lloyd. "Accurate Adversarial Play." *Phalanx: The Bulletin of Military Operations Research*, Jun. 1986.

Hughes, Wayne P. "Attention!" *Phalanx: The Bulletin of Military Operations Research*, Sep. 1985.

Jane, Fred T. "The Naval War Game." Reprinted from the *Scientific American* in U.S. Naval Institute *Proceedings*, Sep. 1903.

Lawrence, Richard D., Lieutenant General, U.S. Army. "Playing the Game: The Role of War Games and Simulations." *Defense/86*, Jan.–Feb. 1986.

LeBourgeois, Julian J., Vice Admiral, U.S. Navy. "President's Notes." *Naval War College Review*, Spring, 1977.

Little, W. McCarty. "The Strategic Naval War Game or Chart Maneu-ver." U.S. Naval Institute *Proceedings*, Dec. 1912.

McCauley, W. F., Rear Admiral, U.S. Navy. "Fight Smart." *Phalanx: Newsletter of Military Operations Research*, Feb. 1984.

MacGowan, Rodger. "Firing Line." *Fire & Movement*, no. 1, 1976.

McHugh, Francis J. "Eighty Years of War Gaming." *Naval War College Review*, Mar. 1969.

———. "Gaming at the Naval War College." U.S. Naval Institute *Pro-ceedings*, Mar. 1964.

———. "Wargaming and the Navy Electronic Warfare Simulator." *Naval War College Review*, Mar. 1967.

344 Mandel, Robert. "Political Gaming and Foreign Policy Making During Crisis." *World Politics*, Jul. 1977.

Muffling, General von. "Anzeige" (Notice). Unpublished trans. by William Leeson, *Militair Wochenblatt* (Berlin), no. 42, 1824.

Murray, Robert J. "A Warfighting Perspective." U.S. Naval Institute *Proceedings*, Oct. 1983.

"NEWS for the Fleet!" *Naval War College Review*, Feb. 1967.

Niblack, A.P., Lieutenant Commander, U.S. Navy. "The Jane Naval War Game in the *Scientific American.*" U.S. Naval Institute *Proceedings*, Sep. 1903.

Nichols, Bill. "CRT." *The Grenadier*, no. 33, Mar. 1988.

Nicolosi, Anthony S. "The Spirit of McCarty Little." U.S. Naval Institute *Proceedings*, Sep. 1984.

"1987 F&M Readers Poll." *Fire & Movement*, no. 56, 1988.

Nordquest, Kurt. "Sophisticated Jutland." The Avalon Hill *General*, Nov.–Dec. 1974.

Pace, Dr. Dale K. "Scenario Use in Naval System Design." *Naval Engineers Journal*, Jan. 1988.

Patrick, Stephen B. "The Rommel Syndrome or Wargames as a Snare and Delusion." *Moves*, no. 1, Feb. 1972.

Reed, Randall C. "Game Reviewing Biases." *Fire & Movement*, no. 42, Winter, 1984.

"Reisswitz the Elder." Unpublished trans. by William Leeson, *Militair Wochenblatt*, no. 73, 1874.

Saha, Mark. "Close Up: *Tobruk.*" *Fire & Movement*, no. 1, 1976.

Selover, Dr. Jay C. "Firing Line." *Fire & Movement*, no. 56, 1988.

Simonsen, Redmond A. "Walter Mitty Strikes Back! or Just Call Me Erwin." *Moves*, no. 1, Feb. 1972.

———, and Dave Robertson. "SpiBus: SPI's Newsletter of Microcomputer/Conflict Simulations Applications, Nr. 0001." *Moves*, no. 44, Apr.–May 1979.

Southard, Jon. "Introductory Games: Problems and Analysis." *Fire & Movement*, no. 51, Aug.–Sep. 1988.

Stone, Bill. "Game Designers' Census." *The Grenadier*, no. 10, Jul. 1980.

Thompson, Frederick D. "Beyond the War Game Mystique." U.S. Naval Institute *Proceedings*, Oct. 1983.

———. "Did We Learn Anything from that Exercise? Could We?" *Naval War College Review*, Jul.–Aug. 1982.

Vannoy, Allan R. "JANUS: The U.S. Army's Tactical Warfare Computer Simulation System." *The Wargamer*, vol. 2, no. 7, Aug. 1988.

Vickery, Colonel John M. "Training for Today and Tomorrow. The Warrior Preparation Center." *Journal of Electronic Defense*, May 1985.

Vlahos, Michael. "Wargaming, an Enforcer of Strategic Realism: 1919–1942." *Naval War College Review*, Mar.–Apr. 1986.

von Reisswitz, Lieutenant G.H.R.J. "Anzeige" (Notice). Unpublished **345**
  trans. by William Leeson, *Militair Wochenblatt* (Berlin), no. 42,
  1824.
"The War Game and How It Is Played." *Scientific American*, 5 Dec.
  1914.

INDEX

356 Military planning process, 68
Miller, Marc, 137
Miniatures wargaming, 127–28,
  142, 207, 226
Mini-games, 140
Mirror imaging, 257
Missile Attack (game), 151
Models: Naval War College places
  in perspective, 8; results as inputs to games, 11; John Clerk's,
  19–21; H. G. Wells's, 34–35;
  Fred T. Jane's, 37; and Navy
  War Games Program, 82; in
  the WARS program, 86; hierarchy of, in NWGS, 92; during the Vietnam era, 123–27;
  professionals borrow from
  hobby, 148–50; and personal
  computers, 151–54, 304; role
  in wargames, 165–66, 181; integrating in design process,
  212–19; in Naval War College
  BDA, 219–21; and development, 232–42; validity of,
  236–37, 266–67; in game
  documentation, 268–69; in
  operations research, 274–76;
  in campaign analyses and wargames, 284
Modern Aids to Planning Program
  (MAPP), 159
Moltke, Field Marshal Helmuth
  von, 30–32
Monarch Services (parent company
  of Avalon Hill), 117
Monopoly, 19, 247, 255
Monster games, 136–38, 143–44
Montgomery, Field Marshal Sir
  Bernard Law, 53
Monty's D-Day (game), 144
Morale, 130, 142, 186
Morison, Samuel Eliot, 287
Morse, Philip M., 106
Mosby's Raiders (game), 175
Move: in Reisswitz game, defined,
    28; defined by the game direc-
    tor, 72; in pol-mil games, 113;

in hobby, 117; simultaneous,
  143; in game structure, 175–
  77; related to time, 222–24; in
  H. G. Wells's game, 231
Moves (magazine), 130, 133, 135,
  144, 151, 229
Move sheet, 252
Muffling, Field Marshal von,
  26, 27
Multi-player gaming, 138
Multi-sided games, 174
Murray, Robert J.: appointed to
  head the Center For Naval
  Warfare Studies, 98; assessment of difficulties at the Center for War Gaming, 99; insti-
  tutes changes at the Center for
  War Gaming, 100; integrates
  wargaming with other tools,
  100; contributions to Naval
  War College gaming, 102; on
  dangers of wargaming, 102;
  mentioned, 103

Naar, Professor Jacques (first holder
  of McCarty Little chair), 89
Nagano, Admiral Osami (chief
  of Japanese Naval General
  Staff), 45
Napoleon (game), 143
Napoleon's Last Battles (game), 140
Napoleon's Last Triumph (game),
  240
National Defense University, 148,
  174, 316
Natural game systems, 315
Naval Tactical Data System, 308
Naval War College: leading role in
  use of wargaming, 8; emphasis
  on human players, 8; similarity of game rules to those of
  Jane, 38; established, 62; Mc-
  Carty Little's games, 66; in-
  sights from McCarty Little's
  gaming, 67; transition to the
  twentieth century, 68; assumes
  major role in navy planning

ing acceptance, 31; utility of, 34, 128; establishment as a hobby, 36; naval, 38; effects of reaction to real wars, 39; strategic, 40; slow rise of board gaming, 40; as a research tool, 41; use of, by Schlieffen, 41; use of, in preparation of German Army manual of operations, 42; German use during World War II, 44; Japanese involvement in, 45; adopted by the Russians, 49; use of, in Great Britain, 50; failure to attract British Army, 51; introduction to the Royal Navy, 51; influence on British planning before World War I, 51; British use in World War II, 53; introduction in the United States, 54; relationship to operations research, 54, 84, 106–7, 154, 155; American use between the wars, 58; as a training aid, 58, 154; use of, by the United States in World War II, 59; importance to the Naval War College, 64–65; integrated with student problem at the Naval War College, 65; value of, 68, 103; limitations of, 69, 83, 127, 157; as an educational device, 70; move to electronic games, 78; Naval War College course in, for fleet officers, 82; the navy's postwar interest in, 84; at the Naval War College, under Stansfield Turner, 89; role in the TCRP, 95; integrating with other tools, 100–101; dangers of, 102; importance of people to, 102–3, 126; contribution of Robert J. Murray to, 102; secret of Naval War College success at, 103; disregarded by operations researchers, 106, 108, 109; use of,

in mining campaign against Japan, 107; and real war, 127; development by the Dunnigan-Simonsen team, 130; use of, to educate civilians about military affairs, 131; in the seventies, 135–36; resurgence of popularity as a research tool, 147; hobby contribution to professional, 148; reintroduced at the Army War College, 148; pushed by General Meyer, 149; beginnings of electronic hobby games, 150; as an analytical tool, 154; postwar resurgence of Japanese interest in, 156; use of, by Israel, 156; Soviet involvement in, 156; in Australia, 158; role of competition in, 158; SASC hearing about, 1, 159; as a discipline for professionals, 159; cyclical nature of popularity of, 159–60, 317, 320; organizing, exploratory, and explanatory roles, 180; difficulties simulating communications, 251; difficulties simulating staff and command structure, 252; historical analysis as a paradigm of, 276; comparison to campaign analysis, 284; key to realism in, 303; simulating the flow of information and time in, 303

War Gaming Department, Naval War College, 82–83
War in Europe (game), 137–38
War in Russia (game), 152
War in the East (game), 137–38, 151
War in the Pacific (game), 137
War in the West (game), 137
Warrior Preparation Center, 296
WARS (Warfare Analysis and Research System), 85–86, 97
Waterloo (game), 199

The Naval Institute Press is the book-publishing arm of the U.S. Naval Institute, a private, nonprofit, membership society for sea service professionals and others who share an interest in naval and maritime affairs. Established in 1873 at the U.S. Naval Academy in Annapolis, Maryland, where its offices remain today, the Naval Institute has members worldwide.

Members of the Naval Institute support the education programs of the society and receive the influential monthly magazine *Proceedings* and discounts on fine nautical prints and on ship and aircraft photos. They also have access to the transcripts of the Institute's Oral History Program and get discounted admission to any of the Institute-sponsored seminars offered around the country.

The Naval Institute also publishes *Naval History* magazine. This colorful bimonthly is filled with entertaining and thought-provoking articles, first-person reminiscences, and dramatic art and photography. Members receive a discount on *Naval History* subscriptions.

The Naval Institute's book-publishing program, begun in 1898 with basic guides to naval practices, has broadened its scope in recent years to include books of more general interest. Now the Naval Institute Press publishes about one hundred titles each year, ranging from how-to books on boating and navigation to battle histories, biographies, ship and aircraft guides, and novels. Institute members receive discounts of 20 to 50 percent on the Press's more than eight hundred books in print.

Full-time students are eligible for special half-price membership rates. Life memberships are also available.

For a free catalog describing Naval Institute Press books currently available, and for further information about subscribing to *Naval History* magazine or about joining the U.S. Naval Institute, please write to:

Membership Department
U.S. Naval Institute
291 Wood Road
Annapolis, MD 21402-5034
Telephone: (800) 233-8764
Fax: (410) 269-7940
Web address: www.usni.org